COSMIC FISHING

COSMIC

FISHING

An Account of Writing *Synergetics*
with Buckminster Fuller

by E. J. APPLEWHITE

MACMILLAN PUBLISHING CO., INC.
NEW YORK

COLLIER MACMILLAN PUBLISHERS
LONDON

Macmillan Publishing Co., Inc.
866 Third Avenue, New York, N.Y. 10022
Collier Macmillan Canada, Ltd.

Library of Congress Cataloging in Publication Data

Applewhite, E J
   Cosmic fishing.

     1. Fuller, Richard Buckminster, 1895–
Synergetics. 2. Fuller, Richard Buckminster,
1895–   I. Title.
Q295.A66   191   76-57203
ISBN 0-02-502710-7

First Printing 1977

Printed in the United States of America

For J Z A

Contents

P R O E M :

"*I care not whether my work be read now or by posterity. I can afford to wait a century for readers when God himself has waited six thousand years for an observer. I triumph. I have stolen the golden secret of the Egyptians. I will indulge my sacred fury.*"

—Johannes Kepler, quoted by
Edgar Allan Poe in *Eureka* (1848)

Introduction

THERE IS no more endearing fact about that
remarkable man Bucky Fuller than his liking for
this book—were he a normal celebrity he'd loathe it
—and no more engaging feature of the book than its
record of Bucky's dudgeon when a detail displeased
him. Naturally details displeased him. Things that
don't get told in public in a man's lifetime, unless
sometimes vengefully, are divulged with a graceful
candor, and not only that, they are shown to the
subject for comment and clarification. It's as if we
had Johnson's comments on a portion of Boswell's
Life. Or no, it's not; Johnson would have roared,
"No, Sir, you are a blockheaded puppy and there's
an end on't," whereas Fuller somewhat frostily offers
to clarify what he takes to be misunderstanding
attributable to haste. He is, *too,* right up to his wife
Anne's stature, near enough anyway that it's
erroneous to describe her as "tall," and besides she
wears high heels. He is *not* an idle yachtsman, which
the author never said he was, but some people may
think it when they see the word "yacht," and what's
more he can reduce time spent on the good ship
Intuition to a mere sliver of the time he spends
working. As for the interpolation on page 90,
linking as it does intellectual and vesical pressure,
it amounts to a hydraulic theory of genius, worthy of
The Tale of a Tub save that it's advanced not Swift-
like, slantwise, but with an Emerson's whole-souled
fervor (though fancy a comparable avowal from
Emerson!).

Emerson's is the name to invoke. Fuller's great-
aunt was Emerson's friend, the redoubtable Margaret

Fuller, whose chair he enshrines in his Philadelphia
apartment, to be sat in for special bouts of
concentration. Like Emerson, he keeps his antennae
poked into the cosmos, where they never cease to
intercept orderly vibrations to which uninstrumented
mortals are deaf. Like Emerson, he expects the most
of all of us; specifically he expects that we will
want to concentrate and go on concentrating, in fact
pay sustained attention for hours on end. And he
posits, like Emerson, that what we deserve from a
sage is austere reassurance, of transmundane—not
to say extragalactic—origin. And they both assert
faith in technology, which they see as patterned
human ingenuity, part of the cosmos because humans
are part of it.

Fuller's vehicle, though, is idiosyncratic mathematics, whereas Emerson's was idiosyncratic
theology. Any graduate of the Harvard Divinity
School could have told you Emerson wasn't a real
theologian, which is just what they'll tell you at the
Institute for Advanced Study if you ask about
Fuller's math. Theology, though, is a tractable
medium for the sage, mathematics (despite the
example of Pythagoras) less so. Emerson's public
read sermons more readily than Fuller's reads the
Clerk Maxwell equations. Not only that, a long-time
osmosis between linguistic domains has left
puddles of soul-talk lying everywhere for non-
specialists to dip into. "Soul" is a word you can use
more freely than "tetrahedron" despite the fact
that few readers will understand "soul" as did
Tillich, let alone Thomas Aquinas, and an audience
that heard Emerson say "oversoul" was halfway to
comprehension before the explanations even started.
On the other hand, when Fuller tells us that the
minimum system, including the minimum thought,

is tetrahedral, he's accomplished nothing until he also explains what a tetrahedron is.

So he has had intuitive recourse to the central tradition of American letters, the tradition of the how-to manual. This is our great indigenous form. From *Walden* (how to build a cabin, know a pond) and *Moby Dick* (how to hunt whales) through *The Sun Also Rises* (how to know Paris, catch fish, fight bulls) to *The Waste Land* (how to understand the masterpieces) and *Zen and the Art of Motorcycle Maintenance* (how to meditate on instruction books), the instruction book has been the armature on which American writers have clapped their philosophic clay. Accordingly, Fuller's "operational mathematics" teaches us to make the fundamental discoveries for ourselves: tying knots, cutting out circles, assembling the three-great-circle octahedron, feeling a tensegrity flex beneath our touch. So much is the reader meant to collaborate, that the famous Moebius-strip marathon Fuller sentences become occasions for collaboration too, grotesque till we cruise vigilantly through them, nimble at the brakes, dipping our headlights at junctions, alert, as wonders flash by, to the glimpsed panorama where range after range of ordered terrain unfolds.

Nor was Fuller's ego affronted by the fact that to get his Sacred Book written might require a collaborator too. "Bucky's *Principia*," remarked one reader, exactly midway between amusement and awe. Newton's *Principia* was one jealous man's grudgingly disclosed system, his most intense concern with receiving credit for his discoveries, not with imparting them. Though Fuller, too, despite some disingenuous candor, accords a high priority to what he calls "accreditation," he is also a compulsive discloser who feasts on another person's signs of comprehension. Deprived of that, he starts talking to himself and soon gets distracted. There was no way he'd write a monstrous big book without another person, so to speak, in the room, and *Synergetics* seems never to have been envisaged except as a collaboration. But various collaborators whose roles hadn't been well defined were quickly chewed up, and it looked as if the book wouldn't get done in Bucky's lifetime until Ed Applewhite entered the project, and Bucky acquired a co-worker in many ways as remarkable as himself.

The reader of *Cosmic Fishing* will soon guess that the experience Ed Applewhite brought to the collaboration was gathered in intelligence work, where it is elementary prudence not to go to another man's island without your own boat. The one time he slipped up, the island was literal, Bear Island off the Maine coast. But to metaphorical islands he came scrupulously prepared, and Fuller, a long-time inadvertent people-eater, never got him cornered. Another intelligence maxim is that you can't gather too much data, and another is that you're most effective sitting still, keeping cool, letting your opposite number make the moves. A temperament that found such precepts congenial was just right for working with Bucky, who needed to write every word of the book himself—if he hadn't it'd be a pointless paste-up, so personal is his vision—but also needed to have some consistency forced upon his tendency to leave last night's brainstorm incompatible with last year's recycled lecture. So Applewhite patiently gathered and cross-indexed 22,000 extracts of Fullerian dicta on every conceivable topic—and many inconceivable—out of which he could usually, as he puts it, "confront Bucky with himself." Fuller, who

generates utterance at the rate of 7,000 words
per hour and has been doing it much of most days
for the past half-century, could thus be contained
at will in a kind of mirror-house, and persuaded
either that he'd already said today's say a good deal
better, or that on the topic of the moment he
generally said something quite different. Such
confrontations never stopped him and weren't
meant to: in a mood somewhere between exasperation and ecstasy he would summon his full powers
to trample contradictions into an ultimate synthesis,
thus achieving some of the big book's most dazzling
passages.

With the undercover man's predilection for
invisibility, Applewhite was also quite willing to let
Bucky mistake him, much of the time, for a typist,
an ideal cover since he does type fast and accurately
and the book had to be typed anyway by someone
who could read not only the author's scrawl, even
when it happened to be on a paper napkin, but also
his multicolored marginalia and the marginalia to
the marginalia, invariably crowded onto the one
sheet however much the expansions might out-
number the original words because Fuller finds the
pressure of limits stimulating (see again page 90).

He is also stimulated by clean print on a clean
page, and a secret reason for Applewhite's care to
keep the triple-spaced foolscap drafts immaculate
was to tempt Bucky into messing them up. This
was partly a way of getting him to clarify the
over-compressed or the hopelessly obscure and partly
a quixotic effort to absorb as much of his
revisionist fervor as possible before printer's proof
offered him more expensive temptations. The latter
effort, to no one's surprise, failed miserably, and
after the whole book had been typeset once it had

to be reset from beginning to end at the author's expense, the work of union operators at their multi-million-dollar machinery having been treated as just one more interim draft.

During most of the *Synergetics* period Applewhite wore a moustache. Subsequently, he confused acquaintances by doffing it, and set a bland new face away from tandem-work with Bucky toward original composition. *Cosmic Fishing* makes manifest what correspondents have long guessed, that Ed Applewhite is a born writer, was in fact a writer all along though he pretended to be a lot of other things. In finally blowing his cover he has given us a privileged, endearing look at one of our natural resources and written the unique inside narrative of how a strange world of words was conquered. If only Herman Melville had had such an aide!

HUGH KENNER

COSMIC FISHING

1. Argument of the Book

ON MONDAY, 20 October 1969, I was sitting at
Buckminster Fuller's desk in his second-floor office
(conveniently above a travel agency) in Carbondale,
Illinois, when a college girl who helped his
secretary with typing brought in a stack of letters
for him to sign when he came to the office. She
had spent much of the night retyping the final clean
versions which he had rewritten many times since
first dictating them. Some were in the fifth or sixth
draft, the earlier versions covered with elaborate
handwritten additions, outlined as balloons or
wedges in the margin, inked sausages running down
to the bottom and up to the top again. There are
very few scratch-outs: no canceling of first thoughts,
just additions of new thoughts. Most of the write-ins
are expanding modifiers, refined nuances of
meaning or apposition. The sentences grow
organically as Fuller crowds on more sail. Dependent
clauses overflow into the margins and seep down
the stairs again like the flood on the Sorcerer's
Apprentice—all without violence to the original
syntax, which survives intact despite the burden of
additional compounds and complexes.

What Fuller loved best as Distinguished University
Professor at Southern Illinois University was
lecturing to large and youthful audiences: this was
his preferred method of teaching. The profuse
marginal outbursts of Fuller's letters and manuscript
pages are the graphic counterparts of his spontaneous thinking-out-loud manner of lecturing,
except that the written page affords a literal record
of his actual thought processes. His style is peculiar

because, while the pages are flat, his thinking is—as he says—omnidirectional, like blowing up a balloon.

The typist feels put upon; she tells me, "Every time I get them all neat, Bucky messes them up again. Why can't he say it all the first time? But I've got it licked now. I've typed these almost to the edge of the paper and he won't have any room. That ought to fix him."

Of course it didn't; he just got a finer pen. I knew it was a lost cause. The margins were often the best part, the icing on the cake.

No one in the office hesitated to call him Bucky, but I suppose few would have had the temerity to sit at his desk and read through his mail. I had known him since I was fifteen—his first teen-age disciple. My family called me "Sonny" and that's what he and his family call me, too. Whenever I have visited him over the years, he has shared with me his wonderfully wide-ranging correspondence, everything from girlish fan letters on pastel writing paper to cranks, mendicants, Montessori teachers, Jesuits, and papers from Nobel medalists. So I sat at his desk reading his mail and waiting to talk to him about writing *Synergetics*, a project which, as it turned out, was to become the exclusive preoccupation of the next five years of my life.

The enigma of this man first emerges when people ask: Who is Buckminster Fuller? They may know that he has written a dozen books—or even that he has the longest entry in *Who's Who*. But what is it that he *does*? What does he profess? What does he do for a living? Is he an architect or an inventor? A teacher or a poet? Engineer or artist? The strategy of his life and his creative energies embrace all these fields and many more. He confounds us at the outset by refusing to be labeled, by affronting our natural propensity for placing him in a familiar pigeonhole. How can we deal with a man who evades every attempt to stamp him with a single tag ending in an *ist* or an *er* or an *or*?

If you ask him, he will explain that he is really a comprehensivist. (He likes to use terms that defy quick reflex and dismissal.) His most certain identity—and his most beguiling self-description— is that he is a "terrific bundle of experience." It is no use asking whether he is a tinker, tailor, merchant, or sailor; if we insist on a category, he will force us to create a new one—and we would be hard put to find a single contemporary to share the label with him. We end up exposing our predilection for what he calls "categoryitis." Perhaps this was what he had in mind all along; like all good teachers he has a flair for gently letting us know when we have asked the wrong question.

Fuller has always been controversial. Shortly after World War II, at a meeting of the editorial board of the *American Scholar* (the journal of Phi Beta Kappa), the astronomer Harlow Shapley proposed that they invite Fuller to write an article for them. Shapley had said, "I suppose Bucky's the brightest man alive." According to the presiding editor, Hiram Hadyn, who is our authority for what occurred,* the proposal was defeated in a chorus of protests in which Fuller was dismissed as a "crackpot" and "eccentric." Shapley said so were Jesus, Galileo, and Einstein, but the rejoinder was

*American Scholar, Spring 1966, p. 190.

unavailing until two decades later when the
American Scholar got around to printing Fuller at
some length.

For me, the dilemma of placing Fuller in the
proper niche, of establishing his role in the world,
was a question of little moment. (Even had I the
qualifications or urge to serve as his apologist, he
would disdain the function.) What was it then that
brought me to Carbondale on that October morning
in 1969 to embark on a five-year commitment
to help him write a single book? . . . a project of
intimate collaboration, symbiotic interdependence,
fine-tuned articulation between the two of us, and
painstaking labor. I think the answer can be put
in one word: curiosity.

All my adult life I had been—as in fact I remain
—insatiably curious about a single aspect of
Fuller's work: his philosophical geometry or his
geometrical philosophy—you could describe it
either way. Of course, he does not use the word
philosophy, because that invokes again the
pernicious pigeonhole notion of an academic specialty, an arcane field of study with special rites
of admission, the misleading assumption of mental
compartments. Instead of philosopher, he says
"thinker." He feels that a Department of Philosophy
is a particular absurdity because it presumes to
preside over a faculty to which every man has
innate access without benefit of academic license.
Thus the subtitle of his eventual book became "The
Geometry of Thinking," expressing the unconventional—even outrageous—proposition that our
thoughts have shape.

Cartoonists often draw thoughts as clouds or
balloons or even light bulbs, with a curving line to
the thinker's skull. But for Fuller, the thinking

process is not a matter of putting anything into the
brain or taking anything out; he defines thinking
as the dismissal of irrelevancies, as the definition
of relationships—relationships that are inevitably
geometrical, and just as inevitably tetrahedral.
(The familiar Egyptian pyramid has a square base;
a tetrahedron may be thought of as a pyramid
with a triangular base.) It is his original conviction
that thoughts not only *may* have shape, but that
they *must* have shape.

The essence of Fuller's synergetic geometry is to
advance a single model to describe the shape of
the physical universe, the shape of energy's
behavior, as well as the shape of metaphysical
universe, which is the shape of our thinking. He
had proposed all his life to write a book attempting
to describe all physical and metaphysical experience in terms of the tetrahedron. What I proposed
was to help him complete this task and to discover
whether I would become a convert in the process.

If the notion of measuring all experience in terms
of tetrahedra seems unduly perverse and abstract,
it is really no more so than our familiar and
unquestioned employment of the cube for the
same purpose.

For two thousand years of Western civilization
and for all the achievements of modern science,
the cube has served as the basic model of geometric
and volumetric measurement. The cube has come in
very handy as the basis of the metric system. The
three XYZ coordinates—the height, length, and
width of conventional three-dimensional measurement—are part of our unconscious cultural heritage,
and we tend to identify reality with this intrinsic
cubical way of describing the physical world.
Centimeters, grams, and seconds (the CGS system)

are accounted in linear or square or cubic modules, and the rules all work with sufficiently exquisite accuracy for man to have reached the moon and returned.

Why then isn't Fuller satisfied with the metric system in the face of its towering pragmatic accomplishments? He concedes that the square and the cube do work in their awkward way, but he argues that their adoption as modules was misguided and erroneous because they have nothing to do with nature's own coordinates. Height, length, and width simply do not exist for him independent of the observer. Thus the observer always inadvertently provides the fourth (or tetrahedral!) point of reference. In his synergetics, height, length, and width exist only as aspects of polyhedra.

With the cube and the square the ancient Greek mathematicians entered the world of nature by the wrong door, eschewing the more elegant triangle and tetrahedron which were so easily available and have been so ignored.

Fuller regards the XYZ-CGS-metric coordinates as the accidental result of man's choosing the wrong tools for calculation, spawning irreducible fractions and irrational numbers like *pi*—with unresolved odd numbers to the right of the decimal point. The advent of the computer has meant that the irrational factors are much more easily dealt with, but in so doing the computer further obscures recognition of the XYZ system as an aberration of man and not as a reflection of nature's own most economical coordination, which is in triangles and tetrahedra rather than squares or cubes.

Though the substitution of the tetrahedron for the cube epitomizes Fuller's major claim in his life's work as a philosopher and mathematician, he had

never formally published the entire scheme of his
argument for academic and public scrutiny. The
landscape of his writing was littered with landmines
in which he had encapsulated only obscure clues
to his geometrical formulations. He had produced
many artifacts, but few handbooks. His triangular
and great-circle tactics were incorporated in his
geodesic domes. His tensional integrity structures—
what he calls "tensegrities"—forsake the compres-
sional bonds of conventional engineering. His many
patents in these fields were manifests of his
original intellectual strategies. (His philosophy was
never a rationale for the domes, rather the domes
were an attempt to explain his philosophy.) But
nowhere was there a systematic and exhaustive
exposition of his claim to have discovered no less
than *the coordinate system of nature.*

Fuller claims not only to have discovered nature's
coordinate system—to which all history up to now
has been blind—but to have revealed how
Einstein's relativity and quantum mechanics can be
demonstrated to popular understanding in simple
geometrical models. In his system, the mysterious
fourth dimension is no longer relegated to the unseen
manipulations of abstract mathematics; the fourth
dimension became visible (to him, if not to me) in
his topological accounting. With him, geometry had
become polemical. The physical universe is composed of matter and energy. His new models
promised to make it possible to observe and
measure the forms and energetic behaviors of the
universe without *pi*, fractions, or irrational constants.
Here was an approach quite unlike that found in
all the textbooks and conventions of Western
scientific teaching. A claim at once so naive and
arrogant boggles the mind. If Fuller is right, can

7

everyone else be wrong? Is this not the classic description of paranoia? Or could it be rational after
all. Or self-demonstrably omnirational, as he
would say.

When I left my home in Washington, D.C., to visit
the Fullers in Carbondale, my thoughts were full
of what I would do if I decided to retire at age 50
on the completion of 25 years of government service,
mostly in intelligence work in Washington and
abroad.

My musty college yearbook records my apparent
intention of going into the advertising business after
graduation. But World War II intervened, and I
never got to wear the gray flannel suit—just five
years in navy-blue serge picking up lint. After the
war I went to Washington and joined a small and
obscure organization then called the Central Intelligence Group. I was aware of alternate and perhaps
easier ways of earning a living, but I was insatiably
curious about the world abroad, and what really
won me over was that I couldn't bear for anyone
to have any secrets I didn't know about.

I was less committed than most of my generation
to the foreign policy of the cold war, as I regarded
anti-communism per se as an insufficient strategic
program. In Washington, then, this was almost like
being a skeptic among Jesuits. My sense of detachment was reinforced by a not-generally-shared
perception that many of the functions of sovereignty
were on the verge of obsolescence. It was happily
too late to re-Balkanize the world for anyone's
game of nations. As it turned out, my sustained
curiosity was adequate compensation for what I
might have lacked in political fervor. I have no
regrets. Our military budgets were to continue at

grossly overblown levels, and I felt it reasonable to spend at least a fraction of a percent of those billions in keeping an eye on what the other side was up to.

Some secrets are kept merely because their revelation would be embarrassing, or disappointing, or both. Most of the secret projects in which I was involved were intrinsically interesting. There must be few people who had more secret clearances than I accumulated in a varied career. When I was detailed as intelligence assistant to Secretary of Defense Robert McNamara, I ultimately got access to just about everything in the book, including the ghastly particulars of our own order of battle. The clearances often had meaningless names, striving for, and achieving, banality; others verged on the poetic; my favorite was called *Cosmic*. I did not jade easily; at least it was a long time in coming.

I had raised a family in Germany, Lebanon, and Washington and had approached the limits of a profession that no longer afforded the charms it held out to me as a young man. I wondered if I should now at last embark on the new adventure of working with Bucky in a way that I could never have afforded to do while honoring the obligations to family and the conventions of earning a living in the real world of politics and government.

I was not sure. I knew that any day spent with Bucky was always full of surprises and excitement and great intellectual stimulation and pleasure in the poetry of his communication. But the two of us were so very different: he had at great cost totally freed himself of all the conventional cultural attitudes of which I was such a happy prisoner by both education and temperament.

Fuller claimed people did not understand science

because science—since the advent of thermo-
dynamics—no longer used models and because the
original Greek mathematicians had made the
mistake of opening the wrong door into physical
reality. Well, I didn't understand science and I
didn't understand Fuller either. But I had known him
long enough and intimately enough to be certain
that no man could be more earnest, more devoted,
more impassioned with his absolute conviction that
he had made geometrical discoveries of un-
paralleled significance. In fact, it was in the nature
of his discoveries to dismiss parallels altogether.

Our familiar three-dimensional reality is figured
in parallels and perpendiculars, the XYZ coordinates.
But Fuller always puts "three-dimensional" in
quotation marks, because he says that what we
really mean by physical reality is four-dimensional.
He dismisses all the parallels and perpendiculars
of conventional measurement as just a squinty-eyed,
special-case, funny, Greek way of looking at the
world. He says the trouble was all compounded
when people tried to model dimensions exclusively
with perpendiculars. Three dimensions can be
modeled with perpendiculars in the cube. Four
dimensions can be modeled with equiangularity in
the tetrahedron. What the three axes of the cube do
for three dimensions, the four axes of the tetrahedron do for four dimensions. The tetrahedron
provides for the convergence and divergence of four
centrally-coordinate planes. He says it is erroneous
to describe time as the fourth dimension (which
I had never really understood); he says that all
dimensions require time. This was not at all clear
to me.

I was intrigued at exploring whether Fuller's
geometry could make sense to me; he claims that

it can be easily understood by any normal child of nine. (Part of his point here is that it is even easier to understand his "natural" coordination if you don't have to unlearn the arbitrary systems now taught in our schools.) If I could master his mathematics—or synergetics, as he calls it—I might perhaps go on to make some sense of modern science in the process. My zeal was tempered only by skepticism. In the last analysis, I think what I really found intolerable was the notion that Fuller could assert such sweeping philosophical claims while having only found time to back them up in piecemeal and cryptic expositions. To me, this was the essence of the challenge. Was there a way he could lay it all out from scratch in a way that even I would understand it?

Only a few seconds after Fuller had got to the office his devoted secretary, Naomi Wallace, had brought a cup of steaming tea to his desk. He was wearing his customary plain, gray-oxford suit that barely contained the bundle of physical energy in his stocky frame. Even at his desk he talked with his whole body, his elbows and fingers dancing a counterpoint to the torrent of commands and questions with which he was starting the day. Wasn't I thrilled by the excitement of the past days' mail? People seemed to be really coming into phase with him in their letters. He seemed to be as keyed up and dependent on this constant stream of written communications as any actor is dependent on response from an audience. As he questioned me about myself and my family, his incredibly soft eyes would fix themselves on me through the thick lenses of his glasses. The temples were secured by an elastic band in the back of the head. His large head,

barely fringed with a neat, white stubble, is
prominent and arresting, conveying at once a sense
of great gentleness and great strength, childlike and
patriarchal, cherubic and ancient of days. The pupils
are enormous gray sea anemones; you can't begin to
gauge their depth. His speech is ebullient, utterly
spontaneous, and devoid of cliché—except for
frequent "darlings" and "for heaven sakes." He talks
with total focus and concentration on the listeners
at hand, modifying his discourse as he intuits their
response, drawing on tireless reserves of psychic
energy.

He was in his office only a few days of every
month and the number of projects competing for his
attention ran into the dozens. Including part-time
student volunteers, there were about 30 people
working on his staff. As Bucky and I talked, there
was a constant stream of young people coming and
going in what the uninitiated might have regarded as
interruptions. His office door was always open and
there was always some delegation or another
waiting for their turn in his presence. Michael was
in charge of the research files and craved 15 minutes
to brief Bucky on some exciting new development
in physics or chemistry. Tom was waiting to get
approval of a new agreement on the S.I.U. (Southern
Illinois University) Library's custody of Fuller's
voluminous archives. Shoji Sadao was Fuller's
architectural partner with offices in Boston; he was
calling that morning with questions about a Fuller
dome to be erected at the S.I.U. Edwardsville
campus. Dale needs guidance about map sales and
new reprints of Fuller's speeches. Don wants Bucky
to address a national conference on nutrition in
Washington the next month. Herbert calls about
scheduling a marathon filming of a World Game

Workshop in the New York Studio School. Naomi
wants to know when Bucky can attend board
meetings at Temcor in Los Angeles and Bangor
Punta in New York. A half a dozen of us would tag
along to the Little Brown Jug for an early lunch
after which Bucky was scheduled to talk (for hours)
to Perk's design class.

At the end of the day, Bucky and I drove the
few blocks from his office to his geodesic dome
house on South Forest Street for dinner with Anne.
The house was a microcosm of Fuller's universe;
spherically coordinate, uncompromisingly simple in
design, and at home in its environment. Its scale
and weathered-wood framing were quite in harmony
with the conventional houses with front porches
and side yards that composed the rest of the elm-
shaded neighborhood. As you enter the house the
first impression is the absence of the familiar
four-square cubical framework of rectangular
floors and straight walls. The effect is totally
disorienting to our reflexive assumption that rooms
should be shaped more or less like shoe boxes.
The result is that Fuller as an architect has created
an artifact—like all of his inventions from the
Dymaxion car to the vast dome at Montreal's Expo
'67, an artifact intended to instruct. You cannot
enter the house on South Forest without receiving
a lesson on how we might organize our environment
with spherical and hexagonal economies simply
not available in a structure where all the rooms
have to be cubes. The dome leads our eye in, out,
and around—not up and down like the box.

The interior walls of the house are a complex
of prisms in which the living room shares a high
dome with a curved balcony containing a library.
Wide glass doors open on a hedged backyard. The

13

abstract unfamiliarity of the design is tempered by the coziness and comfort of the family's furnishings: a ladder-back chair from Bucky's great-aunt Margaret Fuller, big blue antique China dogs from Anne's grandfather, a new telescope, an old barometer, a modern metal network sculpture by Ruth Asawa, an old African carved stool, a Japanese electric clock with an airplane on the sweep second hand—here is a very American blend of the innovative and the traditional, modern technology and fine old craftsmanship, a fitting home for the intermittent residence of a man who says that in this jet-age century for the first time everyone's backyard has become the whole earth.

I first met Bucky and his wife Anne Hewlett Fuller after my sister married a cousin of hers—it was sheer happenstance. Anne is a tall, beautiful woman of regal bearing and always impeccably tailored.*

RBF: *That Anne gives the impression of being tall is really a great victory for her. We are both short by 20th-century dimensions. She always said that she was taller than I, but—being only five-feet-four inches to my five-feet-six inches—she managed to appear so only because her slimness was augmented by putting on three inches of heels. That she always appears—as you say—regal is a matter not only of self-carriage, but of her unselfish tranquility as compounded with her utter confidence in the creative integrity of the mystery of life.*

Although fiercely loyal to her idiosyncratic husband, she is quite a sovereign personality in such matters

*Author's Note: When I reviewed this book in manuscript with Fuller his occasional comments revealed a contrasting point of view, and he kindly agreed that I could incorporate them for the record.—E.J.A.

as her independence of literary taste and her kind
but uncompromising judgments about people.
Anyone who works with Bucky is in Anne's debt
for her unsparing generosity in sharing him with
others.

Anne's family was a large tribe, three generations
of which lived in and around Hewlett and Lawrence
on Long Island's south shore. In the 1930s, Bucky
and Anne had a small but high-ceilinged apartment
in Manhattan's East 80s. With them in town was
their only surviving daughter, Allegra, then a teen-
ager in the Dalton School and a talented student of
both dancing and ice-skating. On Long Island, the
family life centered around Martin's Lane, a rambling
sort of manor house near Lawrence, sheltering
Anne's aunt and bachelor uncles all week and easily
accommodating the second and third generations
who flocked out on weekends. (For some reason I
recall that the Sunday ritual usually included lawn
bowls, applejack and ginger ale, and an enormous
roast of rare mutton.) As I remember it, there were
always parties. The Hewletts were talented, versatile,
and articulate; it seemed only natural that they
should be entertained chiefly by each other. One of
Anne's sisters once declared to me that she had no
desire to meet anyone she didn't already know:
this was not snobbery; she was merely defining an
area—new to me—of social self-sufficiency. Food
and drink appeared effortlessly and abundantly. They
never went to bed, or at least stayed up to all hours.
Among them, family songs and family games came
close to an art form.

*RBF: Anne was the eldest of 10 children. The uncles
still lived in the adjoining house, Rock Hall—the
finest 18th-century house on Long Island—that had
sheltered 10 generations of Hewletts. The Hewletts had*

15

*their children much younger; they had 10 generations
in the time it took the Fullers to have eight. Anne's
forebears on both her father's and mother's sides had
been the original European settlers of Long Island.
They had all that land but they never made any money
out of it and just had to sell it for taxes. And when
we got married the New York papers looked up and
found an early governor of Connecticut in my family
so they described the marriage as a wedding of
"mothball aristocracy." The Hewletts were a lovable
family and Anne's father, Monroe—I loved that man to
pieces. He had built his home, Martin's Lane, on the
Rock Hall property—it was formerly Rockaway Hall,
named for the Rockaway Indians who occupied the
western end of Long Island.*

Bucky managed to revel in the high-spirited life at
Martin's Lane; he was as amusing, as sentimental,
and as fun-loving as any of his perhaps more worldly
in-laws, but without ever compromising his more
serious devotion to design, engineering, and the life
of the mind. It must have galled him that the rest of
them never took any of his ideas seriously.

In those days, what Bucky did more than anything
else was talk. This was in the late 30s, before lecturing had become his chief form of self-expression.
His informal soliloquizing with family and friends
would take on a more earnest tone as the hours went
by. I had never heard such abstract and uninter-
ruptible discourse. Where I had thought that the
industrialization of America was drab, sadly
inevitable, and inhuman in scale, he redescribed it
as an exciting and epochal evolutionary adventure.
Where I had thought that geometry was all very
good but irrelevant, he was saying that you could
not explain life without it. Whatever question we
started with, we always ended up with geometrical
models of foldable circles, or closest-packed spheres,
or tetrahedra. Fuller seized on the closest packing of

spheres not as the abstract formulations of the
crystallographers or molecular chemists, but as the
most elemental expression and arrangement of
nature's energetic forces—of matter. In Fuller's
matrix, the direction from the center of a closest-
packed sphere to the centers of its neighbors is 60
degrees, not 90. The simplest arrangement of closest-
packed spheres is the four whose centers define the
tetrahedron. In those long nights when I was young,
I first learned that the tetrahedron was the initial
conceptual reality to which Fuller's entire philosophical career has since been in homage.

What Fuller was articulating then was his absolute
conviction that physical experience cannot be considered separately from our metaphysical experience
and that they both have behaviors described by the
same geometrical models. Energy has shape; thoughts
have shape; conceptuality organizes itself system-
atically in separating the relevant and the irrelevant,
the observer and the observed. The description of
energy quanta and of abstract thought as tetrahedra
opens up a radical shift in our philosophical perception of reality.

There are no 90-degree angles in nature, Bucky
said then as he says now. There are no square
snowflakes, trees, leaves, nuts, fish, or planets. His
lifelong campaign against the instability of the cube
was beginning to get into high gear. Hugh Kenner,
Fuller's biographer, ascribes his devotion to the
tetrahedron as a recognition of stability incarnate, a
nest of principle.

When Fuller first tells people that there are no
right angles in nature the notion strikes them as
plausible—such is the force of the context of his
argument and his tone of conviction. When I tell my
friends there are no right angles in nature they are

highly skeptical. People find it very comfortable to
see the world, to experience reality, in square
modules, and they find it disturbing—indeed
threatening—for anyone to question squareness.
They do not like to hear that rectangularity may not
be innate; they do not want anyone to invalidate
their scorecard of orthonormality. We measure land
as if the earth were flat; we count cords of wood in
square piles and the volume of a balloon in cubic
feet; we even measure a test tube of blood in cubic
centimeters. People yearn to identify the XYZ
coordinates of geometry, the modules of calculus,
and the cubic lattices of crystals with physical
reality—as if Eden had been laid out on graph paper.
My friends are sure that I have been hopelessly
misled. Every time they come across something
square in nature they hasten to reproach me with
the evidence, not realizing that even "artificial"
right angles like an I-beam or the true cross are held
together at the microscopic level only by an inter-
lacing network of triangulated structuring. One
friend insists that I come see the absolute 90-degree
branch of an oak tree on the ninth hole of his golf
course. Every time he plays the ninth he is reminded
that his friend Eddie is a nut and that Fuller can't
possibly be right.

The Hewletts were prone to conspicuous skepticism about Bucky's unfamiliar way of looking at
the world—such as his substitution of the tetrahedron for the cube. But Anne was staunchly loyal
and shared none of her siblings' misgivings. Her
example—and that of my mother's devotion to
Bucky's bright new picture of the world—made it
easier for me to become perhaps his youngest
disciple at that time. Fuller was always relentlessly

earnest about everyone understanding his message, and though I was just starting off to college, he always treated me—and everyone else—as utter intellectual equals. He describes conversation as the most generous of the arts, and indeed it has always been so with him. I never knew what a tetrahedron was before those summer weekends on Long Island, but I have never forgotten since.

We have seen the advent of adversary urban planning, adversary architecture, and even adversary theater. In *Synergetics,* with its celebration of the tetrahedron, we seemed to have the advent of an adversary geometry. At least in the process of writing about the tetrahedron and synergetic geometry, exposition and program became inextricable. The tetrahedron was also the model for Fuller's psychology, the relationship of self and otherness. He describes the universe as a "scenario of otherness and self." The intention of the work was to present a mathematical model of reality and to redefine the finite physical world in a way that could only improve the human predicament. Improvement of the scenario was to become our banner Excelsior.

So 33 years after those first summer weekends, I found myself in Carbondale about to quit one long career and casting about for another. In our conversation over dinner, Bucky invited me in his generous way to share his entire domain. He suggested that perhaps I could move from Washington to his S.I.U. headquarters and take over the management of his multifarious business and design projects. (I recall a fragment of dialogue that went something like "Bucky, if you were willing to do the kinds of things I would recommend for you to do, then you could easily make enough profits to more

19

than afford what you would have to pay me. . . . But since profit is not one of your interests, and since it would be completely out of character for you to divest control and compromise the integrity of your work, you had better continue to manage things for yourself.")

There were other noncommercial projects we talked about such as his World Design Science Decade Inventory and his World Game studies on which more willing hands were needed. He said he also had more lecture invitations than he could fill; perhaps I could take substitute assignments on the lecture circuit. (Of course I knew that the notion of anyone's substituting for him on the podium was preposterous—but so are some of Fuller's best conversations.) *Synergetics* remained largely unwritten and certainly unpublished. He suggested that perhaps I could work a little on all these projects at the same time. But as the evening wore on, we agreed that I should try to help him with the book—and only with the book.

Fuller's other books—at that time, nine—contained only hints and glimpses of his philosophical geometry. His attempts to present it all in *Synergetics* had engaged the imagination, if not the contractual services, of half a dozen other collaborators before me, people of conspicuous talent, designers, artists, or scholars in their own right.*

Despite documentary evidence of intense effort and industry, these earlier attempts at collaboration had all fallen short of fruition. Perhaps their talents

---

*Notable among these were Duncan Stuart and Peter Pearce. I have sensed their company in the path that I have trod, and they may well have cleared some hurdles on which I have stumbled. As I have neither met nor corresponded with either of them, I cannot report on their experiences in this account.

were too conspicuous. I wondered whether an obscure layman like myself might succeed where they had failed. My chief credential for the project was my profound curiosity about the metaphysical system which he had been hinting at and partially describing for decades, and yet never really presenting in any comprehensive way. I had long been frustrated by this paradox. Our writing together, moreover, seemed the only form of collaboration that would not compel me to move to Carbondale or become unduly a hostage to his unremittingly itinerant lifestyle.

The geometric concepts of *Synergetics* had been gestating in Fuller's mind since his days at the U.S. Naval Academy in Annapolis in 1917, but the original written disclosure, the first documentation, dates from the period during World War II when he was working for the Board of Economic Warfare in Washington. It was in the form of a blind letter dated 14 March 1944, signed and copyrighted by Fuller with nine pages of text and seven pages of diagrams, and captioned "Dymaxion Comprehensive System: Introducing Energetic Geometry." All the essential elements were set down in this enigmatic memorandum, an intensely concentrated blend of prose and symbolic exposition, quirky, dense, and cryptic. Here was the moment of Eureka. Here was the implicit revelation, the geometric lineaments and paradigms later to be made explicit in greater detail in *Synergetics*. Carlson had not then given us the Xerox, and Fuller had 200 blueprint copies made to send to scientists and libraries around the world. My copy was inscribed "To Sonny Applewhite: This has many unedited errors but was a first sticking out of the neck. Bucky."

If Fuller's genuine intention was to convey the nature of his geometrical discoveries to his contemporaries, his strategy with this document was unsuccessful. I have yet to learn of anyone who understands that paper. I have always suspected that its chief function was to establish the priorities of Fuller's claimed discoveries in a kind of private memorandum to posterity—a Rosetta stone of geometries and numbers by which the true significance of his proliferating models and artifacts could eventually be deciphered. The paper attempts to reexamine elementary geometry in the light of Einstein's relativity. He accuses the Greek geometricians of fashioning a cubic block which could not exist in the reality of physical time as described by Einstein. He charges Euclid with a "double foot fault on his first service" by failing to accredit the surface upon which he was inscribing. It is an extraordinary and provocative paper which still awaits evaluation by competent authority.

In Bucky's hypothetical system he treats geometry and number as identical and congruent aspects of all physical phenomena. I had to take this on faith. I recalled that Admiral Byrd had once kept a journal of his thoughts while in isolation at the South Pole. One of his characteristic aperçus was what a great thing it would be for all of us when they get around to setting poetry to music. My own appreciation of the relationship of geometry to number was at a comparable level of innocence.

In a lecture at the University of Oregon on 6 July 1962—one of a series of lectures on nine consecutive days whose transcripts provided a substantial source of expository text for *Synergetics*—Fuller explained to the students:

22

I was urged by a life fellow in chemistry at Oxford University, Sir John Wolfenden, to publish what I called my energetic-synergetic geometry and I published 200 copies of it. . . . He said you will get no credit for this whatsoever and it ought to be in the record because everybody is working secretly [This was in the atmosphere of the intense security surrounding the development of the atomic bomb and the Manhattan Project in the midst of World War II.] and they will use what you have. If you could only be accredited in the other things you are saying it might be more valuable. And yet you are a rank outsider: You are not a physicist or chemist; you don't belong to any society. There are no papers you can publish. . . . Then I published the paper and sent it by registered mail to 200 scientists and that gave me the right to copyright it in the Library of Congress. And so I did.

Inadvertently perhaps, Wolfenden provided not a bad description of the function of a poet: You are a rank outsider. You don't belong.

Since that letter of 1944, a manuscript of sorts had accumulated erratically, partly as lecture transcripts and ephemeral writings, partly the work of other hands. This collection of putative chapters was sequestered in a handsome, black-leather briefcase with the initials "B.F." stamped in silver. It was protected in close custody, never sent by mail, never checked when traveling—but curiously neglected as undone homework. It was called "Energetic-Synergetic Geometry," abbreviated as "En-Syn-Geom," and so pronounced. Fuller has a penchant for abbreviations and initial capitals.

One of Fuller's great close friends in the 1930s and 40s was Thornton Wilder, who was an amateur mathematician as well as a novelist and playwright. Wilder anticipated that *Synergetics* would be Fuller's major work, but he urged him to defer its writing

23

and publication until after he had reached the full
maturity of his lifetime of geometrical explorations.

RBF: When Wilder was encouraging me about my
mathematics he used Newton as a model. He said, "If
your math is as significant as I think it is, you have
the most important book in the field since Newton's
Principia." He said, "You don't often get a chance to
do a new 'Principia,' so you better hold off until you
are sure you have everything you want in it."

Later, I showed him my "Motion Economics" report
for the Board of Economic Warfare—but it was just
too much for him. My global analysis of the practical
effects of synergetics seemed to him to be a virtual
indictment of humanity at large . . . a statement that
everyone else was crazy, with which Thornton could
not go along. He felt more comfortable with my math
because—even though it was just as radical—in
mathematics you don't say that everyone else is
crazy.

It is, of course, not unusual for such ambitious
projects to have a long period of gestation between
the original idea (Fuller would say "conceptioning")
and ultimate publication. In this case it proved to be
a period of 61 years since the first formulation, 31
years since the first written notes.

During this visit in Illinois in 1969, we did not
hammer out any modus operandi. There was the
manuscript, but Fuller was dissatisfied with it and
seemed to want to start again from scratch. As the
most recent of the long succession of collaborators,
Edwin Schlossberg—then studying English and
physics at Columbia—had been working on the
manuscript that spring and summer; but his work,
too, got laid aside during the family's annual August
vacation at Bear Island, Maine.

Without examination of the manuscript material,

and without further discussion of its deficiencies, it
was agreed that he and I should begin to collaborate
on *Synergetics,* the book that was eventually
published in 1975. My first immediate tasks were
two that I could embark on without taking up
any of Fuller's valuable time. I would try to launch
negotiations with the publishers for a larger advance
on royalties so that he could devote more time to
working on the text. And I would exhaustively screen
all of Fuller's published and unpublished writings,
letters, and tape transcripts to identify, excerpt, and
index his every statement relating to synergetic
geometry.

One of the most puzzling aspects of *Synergetics*
is: Why did there have to be any collaborator at all?
Fuller is a man with poetic gifts of expression,
extravagantly articulate, industrious, and self-
disciplined to the point of compulsion—passionately
dedicated to the importance of putting on paper
whatever of his thoughts he feels may be of benefit
to others. He was accustomed always to having
competent secretarial assistance available on short
notice and at whatever cost. When he worked at
Time and *Fortune* he came to rely on the very bright
young research assistants to verify exact dates and
figures. Why should he ever need anything more than
technical or clerical assistance?

Fuller's most successful books, moreover, were
those in which he had had no assistance whatever.
His most effective writing had been that which he
turned out in longhand when utterly alone with his
thoughts. (His two finest essays of this type are
"Total Thinking" and "Omnidirectional Halo,"
discussed in chapter 5.)

If the completed *Synergetics* could have been the

unfettered product of a single artist, would it not
have been a finer work? My answer is that as the
product of a single genius it would have more closely
approached a work of art. But other answers crowd
in. Most importantly, there would probably not have
been any such book at all. He would simply never
have found the time. And much of what is in the
book now was elicited, to use a word with ulterior
overtones from an earlier calling.

RBF: *About why I never got around to writing the*
book. My data was just so voluminous . . . I don't
think that even you have seen all of it. Just books and
trunks of it. And my traveling has been so great. I
have always felt that my thoughts didn't belong to me.
And here I had entered into one of God's rooms—a
whole treasure house that could really be an enormous
resource for humanity.

All that science does is to find out that the physical
universe is technology—that it is the multioptional
intertransformabilities of the complex of generalized
principles governing the eternal integrity of ceaselessly
regenerative scenario universe.

I thought that Synergetics might allow humanity at
large to discover what its options really are. And I
had that kind of responsibility.

EJA: *And, Bucky, I think that part of the problem is*
that a book—even a big book—is not a natural way
for you to express yourself when you are trying to
deal in such large patterns. I think that for you a book
is no more relevant than, say, a vase. For the Greeks a
vase was very important when they had no other way
to keep their wine and their oil, and a lot of their art
went into craters and vases that are now only just in
museums. Well, for you, a book has only such a
temporary cultural relevance perhaps only for a few
recent centuries, but for you—and thinking about a
thousand years from now—a book is just a temporary
convenience, like the Greek vase.

RBF: *Yes, that's what a book is: it's just a tool. A*

book equals a vase . . . equals a lever . . . equals a
tool, a potentially catalytic tool.

Anyway, nothing could restrain Fuller from
rewriting the book from the moment it was in print.
We can hope that he will continue to restate all
the *Synergetics* themes in ever fresh literary forms.
The major task of detailed geometrical exposition
has been accomplished, and this may make it easier
for him to enlarge the scenario. This would be my
hope. At the present writing, he is restating his
synergetic geometry, both graphically and verbally,
working with Tatyana Grosman in West Islip, Long
Island, in the form of a nursery tale parable called
"Goldilocks" on 44 stone lithographs with captions
and drawings wound on a 36-inch tetrahelical
spindle.

2. No Final Draft

FOLLOWING MY VISIT to Carbondale, Fuller's office began to send me a steady stream of articles, clippings, reprints, letters, drafts, manuscripts, files, cassette tapes, transcripts, and ephemeral writings relating in any way to his geometry and philosophy. This was to become the grist for my card index of extracts and cross-references of synergetic geometry. Earlier versions of the basic *Synergetics* manuscript trickled in, revealing that the book had been conceived of partly as an anthology of previously published works and partly of transcripts of lectures, but principally a series of 100 or so master illustrations, assembled and drafted with the help of various hands over the years.

> RBF: *These were neat drafts the students had made of my blackboard drawings disclosing synergetic geometry in hundreds of university lectures.*

In early 1970, while Bucky was giving a lecture in New York, I met him to take custody of the famous black briefcase containing the latest version of the *Synergetics* manuscript. As soon as I got back to Washington I opened the briefcase and found the first chapter on top of the pile. It was entitled "Brain and Mind" and consisted of some 50 pages of blank verse! I was appalled. I telephoned Tom Turner, an assistant to Fuller in the Carbondale office, and asked whether this was a recent addition to the manuscript. (In those days nothing was ever dated.) He said, "Yes, didn't you know, Bucky is going to rewrite all of *Synergetics* in blank verse." He may well have planned to; he may well yet

do so. But I did not dare raise the question frontally.
The prospect of seeing the whole book in blank
verse could be regarded as an exciting vista or an
utter disaster. I had no trouble making up my mind.
Not only do I prefer Fuller's prose or prose poems
to his blank verse with unjustified right margins,
but I know that verse is inaccessible to many people
and that the potential audience for the work would
be restricted accordingly. (There is a theory that if
only Galileo had said in verse that the earth moves,
the Inquisition might have left him alone.)

The original first chapter, "Brain and Mind," was
an essay in epistemology and cosmology. In his
geometry and philosophy, Fuller is utterly
preoccupied with patterns of energy. He says it is
the "fate of energy in the cosmic scheme to
meander through eternity in persistent regenerative
bliss. . . . Energy is the capacity to rearrange
elemental order." And it is the task of his synergetic
geometry to identify energy with number, so it
seems not only reasonable to him, but essential, to
relate man's thought processes—his way of knowing
—to the syntropic-entropic tidal flows of energy,
the largest patterns in his cosmology.

At the same time Fuller was writing "Brain and
Mind," I knew he was also writing another work on
the same theme: the distinction between brain
function and mind function and what he calls
the "hierarchy of degrees of synergy." He had been
rewriting this second work under the title *Intuition*
for two years—since the launching of his sloop of
that name on 31 July 1968. Fuller not only writes
and rewrites but he writes overlapping and
concurrent—but separate—versions of the same
ideas and arguments. The themes, metaphors, and
much of the expository narrative passages of the

29

"Brain and Mind" first chapter and of the *Intuition* draft were overlapping and closely similar but not identical. This circumstance did not bother Fuller in the least—although to me it was unconventional, unorthodox, and ultimately intolerable, because inconsiderate to the reader.

Parallel and concurrent composition—to the point of self-plagiarization—is an intrinsic element of Fuller's method of literary creation. It is the thought that matters, not the form or the particular piece of paper on which he happens to be recording it. Omnidirectional writing again. The same thoughts and sentences crop up throughout the different verses, letters, and manuscripts on his desk at any given moment, demonstrating a verbal Doppler effect.

Bucky agreed with my argument that the "Brain and Mind" first chapter had an integrity of its own separate from the rest of the *Synergetics* manuscript. I suggested that we delete it from the larger work, combine it with *Intuition,* and bring out the two as quickly as possible in a single book of verse. I think it was more the urgency of his message than any considerations of literary form that persuaded him to adopt this tactic. In fact, he would probably have preferred to see each chapter printed separately as we completed it. His constant preoccupation with the urgency of the message is one of his most endearing traits. For him this is not vanity but just a question of trying to live up to his fundamental responsibilities. The Wedding Guest in "The Rime of the Ancient Mariner" was no more self-tormented. The result was that I helped him rearrange the two pieces for joint presentation, reducing the overlappings and redundancies—to some extent.

On the rare occasions when Fuller stops traveling

or lecturing, there is an instantaneous high-tension conflict among projects competing for his attention: architectural blueprints to be screened, drawings to be drawn, exhibit schemes to be sketched, letters to be answered, patents to be filed, friends' books to be introduced, maps and archives to be reviewed, obligations to family and friends, sloops to be sailed . . .

RBF: *You give the impression that I am a philandering yachtsman. I did have just two weeks of racing and cruising in 1971 but since then I have been able to put out only two or three times a summer.*

so this leaves very few of his waking hours for books and poems to be written.

RBF: *I think it is funny that in enumerating all the office tasks requiring my attention, it never even occurs to you to explain that I also have to meet overhead expenses averaging about $200,000 a year. For me, the game of making money has been totally sublimated. As fast as I get money, I spend it. I don't save any money. The result is that people don't tend to think of me in terms of money.*

My rule is to pay bills instantly whenever money comes in; my secretaries and accountants know these standing instructions and they give me an accounting at the end of every month of all monies coming in and all bills paid.

You must not use the credit of others for your own projects. I learned that idea from Henry Ford. I have absolutely no budget. I carry it all in my head. And by knowing that if I have paid the bills and then excess money comes in, then I know I can support new research.

EJA: *This was an omission but not an oversight. I am all too much aware of your daydream accounting. I just find it all so appalling—like keeping a roof over your head with skyhooks.*

In the five years of collaborating on *Synergetics* we averaged less than two days a month of actually working together, and even those days were inevitably interrupted with lecture engagements and television interviews and the like. (In those same five years, I worked on nothing else seven days a week.)

In genuinely desperate exasperation, Bucky once turned to me and groaned—it *was* a groan—"How can it be, Sonny, how can it be that I always seem to have to be paying someone else in order to get out my own books!" He was galled by the injustice of it and suspected that the inevitability of unwanted collaborators resulted from a conspiracy of publishers and literary agents. I think the only reason he has needed the help of others is that he never finds the time to finish the last draft; it is a question of the demands of competing obligations on his time. I told him in vain that Kant never ventured more than ten miles from Königsberg. It is hard to see how anyone with an itinerary like his (see pages 34–35) can describe himself as a writer.

For our first joint working session on the synergetics book, we had arranged to spend all of the second week of February 1971 in a hotel in Sarasota, Florida. Bucky and the indomitable Anne were visiting old family friends, the Alden Hatches, whom they had known since they were married. For once, and indeed the only time, there was a full week for working with no interruptions or competing engagements. On Bucky's instructions, Hatch had reserved two adjoining double suites, the best in the hotel, on the top floor overlooking sparkling Sarasota Bay. Bucky and I were to work in one room during the day; the other suite was for a retired, highly competent stenographer whose services had been engaged for the full week. If the muse were to fail us it would not be for lack of amenities.

I had brought down the sacred neglected manuscript of *Synergetics* in the famous black briefcase. More important, I had a box full of extensive pages of new draft manuscript, several thousand index cards and extracts of basic definitions and concepts, and a proposed new chapter outline with a whole new strategy for the book. This was the fruit of my first year of labor in virtual isolation. Fuller had seen none of it.

In Sarasota I was to find that it is not just Fuller's writing that is overlapping and concurrent; so is his entire creative strategy. At any moment there are always two or three occasional pieces, articles, prefaces to books of others, and blank verse poems in process. "I work on different books at the same time," he says, "like a painter." The degree of versification of the poems tends to be a function not of original conception but of the number of times they have been rewritten. There are galleys to be sent off to meet magazine deadlines. And there are fleeting, new, not fully formulated poetic themes circling Bucky's mental control tower awaiting clearance for a landing. I found that there were new drafts of "Brain and Mind" and *Intuition* on the top of the pile of papers that Bucky had brought with him; these were pieces that I thought had been completed and submitted to the publisher long since. I had yet to learn that chez Bucky there is no final draft. This backlog of always current work in process was the one thing for which I was insufficiently prepared.

Clearly, there was not even the mildest prospect of focusing Fuller's attention on *Synergetics*—on the *book*—until the frail craft in the landing circle

MARCH, CONT.

Mar. 12	Wed.	11:30 a.m.	Fly from Philadelphia, PA United #719
		1:20 p.m.	Arrive Denver, CO (met)
			Attend 28th Conference on World Affairs University of Colorado, Boulder, CO
			Contact: Howard Higman (303) 443-2211 Ex. 648
			Stay: Higman Residence

Mar. 13	Thurs.	11:50 a.m.	Fly from Denver, CO Western #690
		2:30 p.m.	Arrive Minneapolis, MN (met)
			Fly to Marshall, MN, in private air craft
		8:00 p.m.	Speak at SW Minnesota State University, Marshall, MN
			Contact: John Hintz (507) 537-6218 home 537-6639
			Stay: Campus Guest Apt.

Mar. 14	Fri.	10:00 a.m.	Fly from Minneapolis, MN Northwest #712
		2:12 p.m.	Arrive Miami, FL
			Met by Sandy McNaughton. Pick up Mrs. Fuller Drive to private guest house

Mar. 15-20	Sat.-Thurs.		Stay "Shadow Point", Gumbo Limbo Forest, FL

Mar. 21	Fri.	1:30 a.m.- 2:30 a.m.	Attend Candlelight ceremony to commemorate vernal equinox of 1975
			Contact: Sandy McNaughton (305) 945-7518 661-4764 667-4384

An extract of Fuller's itinerary for two
weeks in March 1975. His typed
schedule is revised twice a month with
copies mailed to family and friends.

MARCH, CONT.

Mar. 22	Sat.	day	Workshop - Earth Day Celebration, Peacock Park, Coconut Grove, FL
		evening	Speak at Earth Day Celebration, Coconut Grove Playhouse
			Contact: Sandy McNaughton

Mar. 23	Sun.	1:30 p.m.	Fly from Miami, FL Eastern #192
		3:46 p.m.	Arrive Washington, D.C. (Nat'l)
		5:05 p.m.	Fly from Washington, D.C. (Nat'l) Allegh. #370
		5:50 p.m.	Arrive Harrisburg, PA (met)
		8:00 p.m.	Speak at Pennsylvania State University, University Park, PA
			dinner follows
			Contact: Bob Stoltfus (814) 865-7973 home 865-5555
			Stay: Nittany Lion Inn on campus

Mar. 24	Mon.	11:00 a.m.	Fly from Harrisburg, PA Allegh. #876
		12:02 p.m.	Arrive Boston, MA
		12:40 p.m.	Fly from Boston, MA Allegh. #805
		1:38 p.m.	Arrive Toronto, Canada (met)
		4:00 p.m.	Speak at Great Lakes Symposium of World Society for Ekistics, Toronto, Canada
			Contact: Alexander Leman (416) 964-1865
			Stay: Four Seasons Sheraton Hotel 123 Queen Street West, Toronto, Ontario

had been brought in by the control tower or launched off again to the publishers. They were stacked up, teeming in Fuller's mind. He would pick up a typescript of 40 to 60 pages and urge me to read it to him out loud; if I made some excuse, then he would just read it out loud to me—so great was his engrossment in the most recent rewrites and revisions. Most of "Brain and Mind" and *Intuition* were completely retyped by the stenographer in the adjoining suite. If Fuller changed as little as a single word or phrase, no further progress could be made until the entire page had been retyped. We started work on a Monday. Fuller has an exquisite sense of when to reassert his initiatives in order to forestall anyone daring to bring up a topic not germane to his preoccupations. We started work on a Monday, but it was Thursday afternoon before I had a chance to even mention *Synergetics*.

When we finally got down to going over the manuscript Bucky did not seem to have any feeling of intimacy with it. The first chapter had been peeled off for the *Intuition* book, but I couldn't get him to even look at the remaining chapters. When I asked him why, he said, "It's just not me." He wanted to start over again from scratch, and that was that. He never once, in Sarasota, looked at the manuscript he had been carrying around all those years, and he has never looked at it since.

The original chapter captions were vague and undefinitive; they were reworkings of lecture transcripts with titles stressing exposition by historical review: Trends to Invisibility; Return to Modelability; Conceptuality of Structure. Those phrases captured some poetic notions but they lacked precision and betrayed a casual sense of organization; they lacked continuity of theme. My own program for the book had a much more systematic imperative.

When I asked Bucky where he wanted to start, he asked what did I mean. I said what do you want the first chapter to be, and I presented a proposed outline of eleven chapters, all of which, as it turned out, survive in the final work.

There was only one chapter I had not anticipated, a new one to be called "Omnitopology." (Topology is a branch of geometry dealing with surface relations, the way figures are related rather than their shape or size. Fuller introduces the nuclear analysis of central angles as well as surface angles, hence omnitopology.) Anyway, that final chapter called "Omnitopology" was to develop into a climactic toccata and fugue, a major war plan revealing the strategy of the balance of the book; most of it was written after the manuscript had gone to Macmillan. I have never gone over Niagara Falls in a barrel, but it must be something like the last few months of publishing a book with Buckminster Fuller.

He was not interested in chapters: he likes words, sentences, and paragraphs, as units of thought. Chapters, even whole *books,* strike him as arbitrary constructs, the unnecessary artistic conventions of people with literary preoccupations quite unattuned to the single geodesic structure of his lifetime oeuvre.

Since he had abandoned the old manuscript outline of the accumulated book to that point, I had to persist in asking where to begin. I know now that as always he wanted to begin—like a spider—in the middle and work out to the circumference, or at the circumference and work in toward the middle.

But I had to continue in my linear questioning. Did he want to start with the whole? To start with the universe—his customary strategy—or with synergy, a definition of terms, and a description of his tools of analysis? He would have preferred to do both at the same time and, left to his own devices, could no doubt have done so in one decade or another.

To my unimaginative linear mind, there was a dilemma whether the first chapter should be called "Synergy" or "Universe." He decided to start with synergy, defining our terms of procedure; but I sensed that the decision was largely a concession to me on an apparently unimportant point of procedure. Two years later, when it was too late, he reproached me for not having started the book with chapter 8, "Operational Mathematics," where the reader would first be instructed to use scissors and paste, cutting out circular foldable pie crusts, and using the dividers and straightedge, but not making the mistakes the Greeks had. This would have had pedagogical advantages, but I thought we were being naive enough as it was.

By the time we had to leave Sarasota, we had not agreed on a working program, or even a next step, and I was left with a greater burden of initiative than I would have preferred. But our affection and tolerance for each other were such that we were exhilarated by the prospect of close collaboration and we resolved to complete the book at all costs, certainly within the next 12 to 18 months.

Bucky hardly glanced at my beautiful index files of extracts on all of his synergetic geometry. He explained that any too-fixed attention to parts and details is antagonistic to the synergetic mental processes in which he had disciplined himself. But he was appreciative of all the work I had done and he showed it off with pride to Anne and the Hatches at a cocktail party. What we mainly accomplished in Sarasota had little to do with the book; after 18 months of collaboration, the first chapter had been effectively combined with his poem about the sloop. The result was *Intuition,* brought out by Doubleday in May 1972.

After our first working session in Florida, Bucky and I said goodbye to each other in the airport at Tampa. "The only thing that's expendable," he told me, "is what we do with our time—all the rest is cumulative."

3. Starting From Scratch

WHEN I WAS ASSIGNED to communications duty
in the Navy in World War II, friends felt sorry that
I had drawn such a routine billet, regarded as
lacking in luster, but, as it turned out, I liked the
work. For one thing, so much of war is just waiting,
so I welcomed the mechanical diversion of codes
and ciphers. I even liked the crude systems of strips
and dials that we had to employ with small vessels
and landing craft—more fun than crossword puzzles.
I was fascinated by the pink pages of the cypher
books supplied by the Royal Navy; they were
bound with lead covers for emergency jettison. Best
of all I liked finding out what the classified messages
—ranging in degree of exclusivity from restricted
to confidential to secret to top secret or even ultra
or "magic"—actually said as line by line would roll
up from my typewriter. The work was largely rote
but entertaining, and I often wondered why the
duty was reserved for officers.

It was strict fleet communications doctrine that we
were not supposed to break messages not addressed
to our ship or command, but in the long night
watches I would scan the logs for messages
intended for higher commands or other ships and,
if we held the right code or channel, I would decode
the message just to see what was going on over the
horizon. This practice was frowned upon as
snooping, reading other people's mail.

A good naval despatch has a great economy
of style and a nice regard for distinction of
subordination and prerogative; the personality of

the commander always comes through the
operational and tactical jargon. The ultimate
satisfaction was recognizing that an urgent or
operational immediate despatch was of sufficient
import for me to take it directly to the captain, or
the admiral, with the certain judgment that I was
correct in having the orderly wake him. I would
then bear back their always blunt and terse replies
or orders with a flourishing "Aye, aye, admiral." (I
was 22 and zealous.)

At the height of the German U-boat campaign
against the lifeline convoys to Britain in 1942, I was
transferred as assistant communications officer to
the then top-priority fleet task force known
officially, even extravagantly, as Commander Aircraft
Antisubmarine Warfare Development Detachment
Atlantic, or more frugally, as ComAirAsDevLant,
and even irreverently as HalfAssDevLant. We
reported directly to Admiral King in Washington.
Our job was to assess the latest tactics and hideous
devices of Admiral Doenitz's U-boats and to
develop strategies and test countermeasures against
them. It was a sophisticated contest with very high
stakes as any new weapon or tactic had an
operational life of only about six to nine months
until the other side caught on and devised defenses
or evasive action against them. We had every kind
of plane, submarine, and patrol craft at our disposal
as well as a concentration of technical talent,
military and civilian. Among the latter were Starling
Burgess (who designed the America's Cup defenders
and had worked with Fuller on his Dymaxion car in
Bridgeport in 1933) designing underwater gear like
torpedo decoys, and Roark Bradford (the southern
novelist) writing crash-deadline training manuals

for the Navy pilots. At that time the convoy losses were almost sinking Britain and we felt it was up to us to save the day.

Every day, fair weather or foul, ComAirAsDevLant staged elaborate drills and exercises in the Atlantic off Newport. After dusk, when the last patrols and planes had returned to base, the exhausted operations chiefs and engineers and technical specialists all assembled for the nightly meeting to assess the day's findings which they were obliged to detail in a summary despatch to Washington. Most of these cables that I had to set up and encode for transmission seemed unduly hasty and prolix. In the isolation of the code room, I started to reorganize the messages and try to simplify them—really just to save time in transmission and processing. This was not strictly the function of an assistant communications officer, but the talented operations and technical officers were overworked and they welcomed any help they could get in what they regarded as their most onerous task: trying to get it all down on paper at the end of the day.

After a week of my re-editing their drafts when they had reached the code room, they asked me to sit in on their sessions while they were still sorting out the instrument read-outs and significance of the day's events. About halfway through each nightly session, when I could first sense a consensus emerging, I would sneak off to my trusty Underwood radio typewriter (it was all set in caps with no lower case; the radiomen called it a "mill") and type a first draft despatch with half a dozen carbons for the chief experts. I knew nothing of mathematics or engineering—not to mention the arcane acoustic and electronic subtleties involved.

42

I found that even with only the dimmest grasp of the matter, I could summon some deep structure of organization and at least get the issues down in some sort of sequence on paper. They always found it easier to work from and correct my first draft, however erroneous, than to start off writing from scratch.

Starting from scratch is the hardest part of writing. So, after my first session with Fuller in Sarasota, it was much in the spirit of those high-pressure critical days in World War II, and dealing with a subject at least as complex and unfamiliar as antisubmarine warfare, I started to see if I could provide him with first drafts of the synergetic geometry. I knew that it would be easier for him to rewrite than to write, and once *Synergetics* was on paper again it would be easier to engage his daily attention and energies.

By this time my abstract files were so extensive that I found I could not only put together the first drafts, but that I had enough material at hand to put most of the first drafts in his own words— often verbatim but seldom seriatim—as most of those words were recorded over a period of decades.

For instance, in Sec. 642.01* the first sentence is from the transcript of a lecture at Eugene, Oregon, of 8 July 1962; the second from a lecture in Carbondale of 21 October 1965; the third from "Omnidirectional Halo" (1959); the fourth was written by Fuller on the manuscript on 2 June 1971; and the fifth is from a lecture to NASA in June 1966. Sec. 440.05 is another example of a paragraph composited from sentences drawn alternately from lecture transcripts: 11 July 1962,

---

*Section references are to *Synergetics*.

June 1966, and 13 November 1969. In cases like
these Fuller invariably adopted the sequence I had
proposed in the first draft and added minor
embellishments and modifiers in the gaping margins.

Two years after we started working together,
Fuller and I finally got around to a written agreement
on the project. We signed a standard ghost writer's
contract in which he granted me a minority
participation in the royalties and subsidiary
rights and I conceded his sole claim to authorship.
Quite aside from the contract and far more
helpful as a guide to a collaboration, we agreed
from the outset on four explicit understandings
or guidelines:

(a) all the geometry is to be expressed in words,
in writing;

(b) I was not to be a co-author;

(c) all text was to be in Fuller's phraseology; and

(d) our writing would make no attempt to
conform to technical textbook or academic
standards.

One of the first questions I asked Fuller was
whether he felt the whole of his synergetic
geometry could be expressed through the written
word. Could it all be explained without pictures,
models, mathematics, or symbols? Over the years
an enormous amount of illustrative and graphic
material had been accumulated, several filing
cabinets of photographs, models, drawings, diagrams,
and charts. Chief among these were Fuller's own
drawings and his own photographs of models he
had made—highly concentrated displays of closest-
sphere-packing geometry, marvelously executed
and densely captioned. (These were all too often

44

executed on the most convenient material at hand—
old shirt boards, ruled legal pads, even laundry lists
and envelopes.) There were also beautiful drawings
by Duncan Stuart and more conventional drafting
by Shoji Sadao, Robert Brooks, and Peter Pearce.

*RBF: Stuart's drawings were redrawings. Brooks and
Pearce were paid for the work, which was to make
clean India-ink versions of my original sketches.*

In fact, the earlier collaborators on synergetics had
contributed mostly in the form of graphics, with only
superficial attention to the text.

*RBF: They made only larger models or cleaner
graphics. Their attention to text was limited to
phrasing and syntax.*

Much of the original text was in the form of
captions to illustrations. Captions are a literary
device in which Fuller delights. In *The Dymaxion
World of Buckminster Fuller,* which he wrote with
Robert Marks, some of the meatiest matter is
presented in the form of captions.*
(This practice was probably exacerbated by Fuller's
wanting to add to the text after the book was well
along in production.)

The set of one hundred or more master
synergetics illustrations was constantly being
rearranged in new sequences, each with its own
rationale; but aside from the captions, there was no
narrative connection or attempt at systematic
exposition other than the disembodied, overlapping,
and not specifically referenced lecture transcripts.
My concern was that so many of the pictures were
neither self-explanatory nor otherwise explained.

*See especially Fuller and Marks, Figs. I1 through I8 in the
original hardcover edition (Reinhold) and Figs. 228 through
235 in paperback (Doubleday Anchor).

There would clearly be no contribution for me to make unless help was needed with the text. I cannot draw. I had studied no mathematics beyond high-school geometry and algebra, which I loved; but I had no stomach for trigonometry as I was—even then anticipating synergetics—uncomfortable with the notion of ratios between angles and edges; this was something I could not conceptualize although I could accept it as a convenience producing useful results. My favorite courses in college were philosophy. I dropped chemistry when everyone in the class except me seemed to understand the concept of valency. Later, in the course of editing *Synergetics,* I reschooled myself with my son's physics textbooks. I did not consider that my mathematical innocence should disqualify me from collaborating on *this kind* of a book on geometry; far from it. Since my mathematics was so rudimentary, I would have less urge to correct my mentor, and I knew that this temptation had been the undoing of some of my predecessors. In my case it took less daring to be naive. What I could do—and did—was to track down every word Fuller had written or said on a given topic, rearrange those words in what seemed an orderly sequence, and thus confront him with every word he had said on the subject.

Fuller assured me that while the pictures could illuminate the text and greatly facilitate comprehension, everything in his geometry could be put into words. This simple proposition epitomized our collaboration. As a discipline, we agreed to try to put the geometry into words in such a way that the total system could be understood if read aloud to a blind man. This had some occasional virtuoso results, like Sec. 506 which, without pictures, describes how to tie a knot. I wonder how many people have tried to do that before.

In adhering to the constraint of trying to put the whole geometry into words, Fuller was conscientious if not exemplary. He has a propensity for tape recordings. One of the troubles with tape recording Fuller's talk is that so much of the message is conveyed in body English—hips twirling imaginary hula skirts of ball bearings, hands describing the coupling of universal joints, elbows pumping precessionally as pistons of an internal combustion engine. Often when he was dictating to me, or rather sharing an articulated stream of consciousness, he would become exasperated when I would not look at what he was doing with his hands. I would have to tell him that this was deliberate, that I was not looking on purpose. I would say we are writing a message to be found in a bottle; the reader won't be able to see your hands. On these occasions he was not so much writing a book as just explaining the math to me. His sole concern seemed to be that I should understand. Anyone who has ever talked to Bucky has had this experience of his intense desire to explain a process for its own sake. I was no different, just the ordinary village listener; he would be oblivious at such times of what should go into the text or of the business of writing a book.

> RBF: *I often would talk to you in sort of personal asides of free association. I had no idea that you were putting everything down. We didn't start off that way or plan it, but it ended up with what I called cosmic fishing, where the intuitive asides become the main account. The effect is precessional.*

> EJA: *Maybe that is why I never like to have a tape recorder around when we are working together.*

Our verbal discipline reached its purest form over the telephone when I really couldn't see his hands. He would often call up whenever an idea came to him or to refine some statement that lingered on from our last session working together. Yet even when calling from overseas, he was always punctilious about never getting me out of bed. Many passages that survive in the book were dictated in this way, not just from Carbondale or Philadelphia—where he later moved—but from La Jolla, Tokyo, and New Delhi as well. The only time he called collect was from Windsor Castle—Fuller is an exemplary house guest.

The second crucial understanding was that I was in no sense to be considered a co-author. Though the final publishing contract eventually described the two of us as co-authors, Fuller accepted my assurance that as far as I was concerned the book had only one author, and he was it. The entire content and substance of the work is his; my chief function was that of eliciting the material, organizing its presentation, and policing it for consistency of style—not matters of essence. Even though I did virtually all of the first drafts, they were derived from what he had first said to me or others, or were developed from material implicit elsewhere in his work. I did suggest some words, such as "epistemography" for Fuller's concept that knowing has shape. In the process of eliciting, I explored certain geometrical relationships and implications which were unfamiliar to me and about which I was curious, but I claim no original proposals or discoveries. Moreover, there are to this day certain passages in this book that I would be hard put to explain to someone else, and for this reason alone I could not assume the responsibility of co-authorship. (I find certain passages sublime, others that verge on madness.)

As the book took shape, it became more and more apparent that it was totally different in form from Fuller's other writing. There was ineluctable evidence of the work of an alien hand. Thus we agreed at the time of completion to put my name on the title page, in smaller letters and with the designation of "editor." After page proofs were completed and the book was ready to go to press, one of the publishers' perennially new editors ruled that my designation as editor would have to be stricken. He said the word "editor" implied that the work was an anthology. I argued in vain that the simplest glance would reveal it was not an anthology. I was sorry he was so adamant, but in retrospect I can see his point as most books with credited editors are indeed anthologies or compendiums of some sort. There were only two options open: I was to be listed as co-author or my name was to be dropped. Either course would involve deceit. After much painful negotiation between Fuller in Maine, our publishers and attorney in New York, and myself in Washington, a compromise was agreed on: the title page would read "in collaboration with E. J. Applewhite," in the same smaller letters as before. This also prompted the drafting of my terse and virtually unqualified disclaimer following the title page. It required the better part of two days and a dozen drafts before Fuller and I finally agreed on the wording of that note. It reads:

The discoveries, concepts, vocabulary, phraseology— every word—and entire writing style of this book originate exclusively in the mind of R. Buckminster Fuller. My role has been strictly editorial: identifying, sorting out, and organizing the presentation of five decades of Fuller's thinking, continually confronting the author with himself.

E.J.A.

That note incorporated the third ground rule of our collaboration, that I would never write a word for Fuller but would present him with his own statements in what seemed a logical arrangement. The first drafts would often contain non sequiturs and apparent contradictions which I would leave verbatim and starkly unresolved, but watching him like a hawk, confident that rewriting would restore consistency. It almost always did, and the effect was totally synergetic. If we have anything at all to contribute to the history or technique of joint composition, it could be summed up in that phrase: confronting the author with himself. That was the law and the prophets.

My growing file of index cards contained extracts, cited by date and source, of every topic I thought appropriate for eventual inclusion in *Synergetics*. Thus, in drafting the definitions that introduce every chapter of the book, I was able to draw upon variations in their restatements over a period of decades. On legal and aesthetic grounds we had agreed to abandon the anthological character of the original manuscript. By breaking the earlier writings into pieces and redigesting them, it was our desire to present their themes without doing violence to their integrity as already published works. Self-plagiarism within the law.

This modus operandi was completely congenial to me; where Fuller is always synthesizing, my temperament is analytic. His whole strategy is that of "starting with Universe," going from wholes to particulars. My reflexes are always trying to identify the parts and give them names so they will be at home in my more conventional cultural landscape. Though the result was synergetic, it was at the price of ending up with a book that Fuller has a little trouble finding his way around in. This is not too much on my conscience. His original conceptions are omnidirectional, but a book is inescapably linear. He has found a way to "unwrap the orbitals" on a linear reel, he says, but it involves substituting a tetrahedron for the IBM golf ball on the typewriter.

The fourth imperative of collaboration was agreement to disregard conventional standards of technical or textbook exposition; I was unfamiliar with them and he abhorred them. Our unfortunate use of "vertexes" instead of the comelier "vertices" is a dead giveaway of our amateur standing; no pukka geometrician would ever employ "vertexes," which I chose simply because it is listed first in the dictionaries.

There was the goal of getting the book correct and the goal of getting the book exactly the way Bucky wanted it. These two goals were often in happy coincidence, but when they were in conflict I had no question about which course should prevail. Fuller's poetic expressions and his standards of measurement often departed widely from conventional descriptions of physical and chemical functions; when they did, the license of the poet was always honored. The result is that Fuller has a remarkable confidence in the integrity of the text as saying things in exactly the way he wanted to say them. (This applies to the wording of the text in contrast to its organization.) Each successive draft was painstakingly dated and documented at the time it was superseded; no pages were cut up or re-pasted. No draft page was ever destroyed.

Every manuscript page was authenticated—where not indeed obscured—by his holography.

There was a final implicit premise of our collaboration that had the greatest potential for discord but seldom resulted in open conflict. For Fuller, the overriding imperative in the book was to achieve perfection. For me, it was simply to get the book out, and in his lifetime. I had no interest in unpublished perfection. He was urgent to get his message into the public domain and into the hands of eager young adherents, but his zeal was tempered by thoughts that once it was all in print and bound between covers there might be loss of control. I was eager to see the bird fly the coop, little suspecting that he would keep it tethered like a falcon.

I knew it was not within the nature of things that I could give Bucky assurance that *Synergetics* would have everything in it just the way he wanted it. His travels were so constant and the manuscript was so voluminous that he had no familiarity with it and no access to it except for those few crowded days a month when we were working together in Washington. Thus he was extremely dependent upon me for the organization of the text and the illustrations. He was particularly pained—but most reasonable and resigned—about having to drop so many illustrations, especially photographs, due to the constraints of budgets and production. The chief promise I made him was that the book would contain no text that he had not, at one time or another, wanted included in that specific form. This commitment was honored. At the same time, I warned him that there was no practical way to insure that everything he wanted in would get in; and of course it didn't.

Within a month of publication he was adding

inserts in the margins of my first copy and has been doing so ever since. For most of us, the act of publishing is irrevocable. But for him, not quite; he has never been deterred from rewriting a book just because it has been published. Not any more than he will stop rewriting a letter just because it has been mailed, or even because the letter was written by someone else to him, or even because the person to whom it was addressed or from whom it came has since died. He often speaks of his mother, and even my mother, in terms revealing that for him the dead are never quite departed. In speaking, as he often does, of I-they and we-me, he creates new burdens for hyphens. This tendency to merge pronouns relates to his psychological geometry compounded with a fractured sense of identity. For him, individual and group consciousness are inseparable aspects of the same phenomenon. The relationship between self and otherness is, of course, tetrahedral. I have the feeling that in the ultimate refinement of his psychological geometry he will arrive at some kind of prismatic Golden Rule—going St. Matthew one better by putting pronouns out of business altogether.

4. A Peculiar Accuracy

AT THE END of World War II, Bucky invited me to join his new firm building mass-produced aluminum dome houses in the Beech Aircraft Company plant in Wichita, Kansas. I had just completed a three-year tour on an aircraft carrier in the Pacific and part of the charm of his offer was that Wichita is within a few miles of the furthermost point from salt water in the United States. Fuller was chief engineer and chairman of the board; I was to become personnel manager. The spectacular failure of this project has been well documented by Alden Hatch,* but it was attracting nationwide attention at a time when both the unions and Wall Street investors were eager to support the conversion of war plants to peacetime production, particularly in a field like housing with its five-year construction backlog. I remember one day being sent out at the last minute to the airport to greet a prospective investor, a man named Giannini; it was only some time later that I learned he was actually A. P. Giannini, the founder and head of the Bank of America.

While Fuller regretted the fiasco, he did not regard the venture as a personal failure. He knew that the dome house as an artifact was ahead of its time, but it was a satisfactory industrial model not just as an end product but for the implicit new distribution and service patterns involved. It was the same with his Dymaxion car in 1934. He still smiles when people tell him they are sorry for him that only three of his three-wheel cars ever got built—that it never got into

*See chapter XIV of *Buckminster Fuller: At Home in the Universe* (New York: Crown Publishers, 1974).

production. But he was only trying to build a proto-
type, he was not trying to get into the automobile
business.

During the Wichita period Fuller had little time
for working on his geometry and mathematics. Of
course, the drawers and shelves at his house and in
his office were full of the familiar closest-sphere-
packing models glued together in a great variety of
pyramids and tetrahedra. But he only got time to
break them out late at night or on weekends. Their
significance to him could be observed, but not
shared, as he was chiefly engrossed in his industrial
philosophy and its engineering applications—what
he calls "design science." During the winter of 1945–
1946, he presented long talks at least weekly,
explaining his industrial strategies to the new
engineers huddled intently over their drafting tables.
The design offices at Fuller Houses, Inc., often took
on a pedagogical character—a sort of technological
Chautauqua—with which these young draftsmen
from Chicago and Detroit were quite unfamiliar.

Though it was not related to my office responsibilities I would record and transcribe these talks.
After half a dozen had accumulated, I prepared a
composite version which the company printed up in
a 42-page pamphlet under the title "Designing a New
Industry."* Fuller did not rewrite or edit the piece
and he barely found time to write a brief introduction. Fuller signed the introduction: my explanatory
note is anonymous.

With my penchant for taking things apart and
analyzing them, I prefaced the pamphlet with an
elaborately detailed topical outline of the composite
lecture with the sequence divided into 12 main head-

*Eventually published as chapter 8 of *The Buckminster Fuller
Reader*, ed. James Meller (London: Jonathan Cape, 1970).

ings with three levels of subordinate description. In preparing this summary version, I attempted to translate Fuller into plain speech by shortening the sentences, avoiding the nonce words, and substituting a more conventional vocabulary. I thought I could do it the way *Reader's Digest* does. But in the process my effort to preserve the spontaneity and personality of the speaker was unsuccessful. The flat statements survive, but all the sound of his essential Bucky-ness is muted. It does not soar; it reads as if it had been written with gloves on and then corrected by a grammar teacher. When I first started working on the synergetic geometry—some 23 years later—I was tempted to try the same pedestrian approach, but I found that it couldn't be done. The strategy of Fuller's idiom is too subtle. If you tinker too much with his sentences, they just turn into pumpkins.

A lesser man might have been discouraged by the Wichita experience, but Fuller's resolve was unshaken and the next several years—spent between Anne and Bucky's apartment in Forest Hills, Long Island, and at Black Mountain College near Asheville, North Carolina—were among the most productive in philosophical and design breakthroughs of his entire career. Less than a year after Fuller Houses went out of business, he threw his energies into compiling a detailed recapitulation of all published references to him and his work—a sort of bibliographical stock-taking that he had printed up as his "Dymaxion Index 1927–1947." He sent me a bound copy at the time, inscribed in his hand on the cover:

To "Uncle" Edgar Applewhite, able custodian of my own incoherent items on this list, who will some day clarify whatever may be worth clarifying in these items.—Buckminster Fuller

The reckless combination of modesty, faith, and prophecy is characteristic.

Fuller's primary vocation is as a poet. All his disciplines and talents—architect, engineer, philosopher, inventor, artist, cartographer, teacher—are just so many aspects of his chief function as integrator. When he was appointed to the Charles Eliot Norton chair of poetry at Harvard in 1962, he described the word "poet" as a very general term for a person who puts things together in an era of great specialization when most people are differentiating or taking things apart. Poetry, for him, is a calling to which the individual may aspire, but which he may not profess; a poet who succeeds in his function may be recognized only by others—and then only in retrospect. For Fuller, the stuff of poetry is the patterns of human behavior and the environment, and the interacting hierarchies of physics and design and industry. This is why he can describe Einstein and Henry Ford as the greatest poets of the 20th century.

> RBF: I am enthusiastic over Emerson's definition of poetry as saying the most important things in the simplest way. You couldn't say anything more important in a simpler way than to say $E = Mc^2$. That is why artists and the young people really go for that equation—even without knowing too much about it. And Ford was an artist because he painted a world-embracing scenario—an interaction of all the resources and all the people.

The language of Fuller's poetry, his diction, deliberately tries to avoid the connotations of our cultural, sentimental, or traditional heritage—or with what other people often mean by poetry. His writing —or at least the aim of his writing as he sees it— has nothing to do with aesthetics and ethics or

artistic and literary traditions. Those are all words
that he does not even use. We were fond of quoting
Barnet Newman: "Aesthetics is for the artist what
ornithology is for the birds." Though Fuller would
reject the classicism of Joyce's allusion, he would
share his view of Daedalus: that the walls of the
labyrinth are religion and nationalism.

Culture is of interest to Fuller only to the extent
that its artifacts are instructive of man's relationship
to his environment in the larger biological or
evolutionary sense. In his lexicon, most of our
common appreciation of history, art, and literature
are just so many highly conditioned—often impeding
—reflexes that do not necessarily represent any net
gains or useful insights (1066 and all that). Rather
than try to learn from the past, we will do better to
try to start ab initio.

Shelley may have been the first to herald the
sweetness of the unheard song. Fuller not only
agrees, but has enlarged the dictum to celebrate the
power of the unseen picture. He is forever reminding
us that 99 percent of physical reality—all of nature's
energy—operates at frequencies too high or too low
to be tuned-in within the limited range of human
visible or audible scanning. So for him, any theory
of art, or poetic function, can be only partial—indeed
fractional—if it treats only within the limits of what
we can see and hear.

If Fuller's ultimate aesthetic is invisible, so is the
music of his poetry virtually silent. There is no song
or rhythm in his verse. There is no rhyme. The
cadence is liturgically repetitive. For him the artists
are philosophers in cry, and thus his verse is
inevitably programmatic and polemical. He writes
primers of new-fashioned Mother Goose tales . . .
like choruses from blueprints. The metaphor is not

so much in phrase as in a pattern of myths retold and of technology made graphic—as in ". . . the echoed voice of a poling raftsman to the forwardly informative radar manipulations of the stratojet-piloting airman."

He avoids the term "20th century" because he regards it as anticosmic, reflecting a myopic and parochial time-accounting derived only from the recent religious history of a minority group of humans on Planet Earth. For him, engineering is poetry, and he says he would not be surprised if some day it were proven a law that the better the science the better the poetry. No wonder his style is peculiar, his vocabulary strange to the ear, his syntax so remote from common expression. It is no accident: in fact it is a great deliberate effort to enter the avenues of the listener's mind without tripping the familiar and comfortable reflexes. How could one better dispel dreams of marble halls—or better epater le bourgeois—than come up with a phrase like "dwelling machines."

Fuller is not only noncultural, he is nonpolitical. He has a manifesto, but it is strictly technological: do not reform man, rather reform his environment. He foresees for us a new world in which houses are not only dwelling machines, but "environment-altering *valves*." Government and politics have nothing to do with it. Politicians are merely accessories after the fact. Socialism is merely "a boring way to speed up the mess." The intent of all his writing is polemical, but in the last analysis it is the artifact that will save the day. Society will only take on the new, he says, when nothing else will work. "You can't better the world by simply talking to it. Philosophy to be effective must be mechanically applied."

I had read somewhere that a similar message could

be found on the grave of Karl Marx. I was skeptical
of this, and on a recent trip to London I went out to
Highgate Cemetery to read the inscription for myself:
"The philosophers have only interpreted the world in
various ways. The point, however, is to change it."
—from *The German Ideology*

In March of 1970, Buckminster Fuller spoke before
a great—and some witnesses say, restless—throng
in the Methodist Central Hall, Westminster, London.
The following week the *Times Literary Supplement*
reported that "the audience seemed numbed both by
his convoluted rightness of detail and by the basic
simplicity of his moral beliefs. . . . Fuller himself is
seldom thought of as having any literary importance,
but he has meditated deeply on language and uses
it most distinctively—not lucidly but with a peculiar
accuracy." I thought the remarks lacked charity and
took umbrage at their condescension of tone, but I
concluded that his "peculiar accuracy" is what an
editor must preserve at all costs.

In referring to his "moral beliefs," the *TLS*
employed a phrase that Fuller—in his striving for
accuracy—always avoids. "I don't believe anything,"
he says. A proposition is either experimentally
demonstrable or it is not. A generalized principle is
either discovered and proven or it is not. He doesn't
need beliefs any more than he needs preachers or
leaders. He considers most of our notions of good or
evil as so many cultural or popular reflexes. He does
not use the term *ethics* and does not tend to think
of a world of good and bad. An electron is not bad
because we give it a negative sign. The world is not
getting any better—better understood, perhaps, but
not better. "*Shame* is related to the words we've
invented, but evolution has her own accounting."*

*RBF to Barry Farrell, Bear Island, 18 August 1970.

His search for this kind of idiosyncratic accuracy in speech is one of long standing. In the 1930s, Fuller was one of the first enthusiasts of I. A. Richards and Basic English, a language purged of cultural nuances if there ever was one.

My conviction that Fuller is untranslatable was reinforced by Reyner Banham's argument in *Arts Magazine,* London, October 1963, that it is almost impossible to write for Fuller or even about him except in his own idiom: this is because "his creative thinking does genuinely seem to be done in the grammar and syntax he uses when speaking. Problems are simultaneously bulldozed frontally, undermined termitically, and outflanked by relative clauses lasting up to six weeks." (I could put italics on "or even about him.")

An earlier draft of Sec. 223.01 had one sentence that ran for four manuscript pages; here was a lawyer's instinct for having all the conditions equally and concurrently qualifying. (Interqualifying, Bucky would say.) His language does reflect the way he thinks and in that strange idiom he declaims to physicists as if they were eight-year-olds—and confides to eight-year-olds as if they were physicists.

In February 1973, we were going through our third draft of Sec. 1009.68 when I looked up and said, "Bucky, you know people are going to wonder what the hell Fuller means with all this talk about 'intertransformabilities.' . . . They're going to say that's just so much horseshit." He agreed instantly. "Yes, that's right. That's just what horseshit is, a beautiful example of intertransformability." Fuller never uses barnyard language except explicitly. He was amused two years later when our friend Bill Marlin, as architectural critic at McGraw-Hill, told us that one of his colleagues, a biophysicist, on first examining *Synergetics* marveled at the book's innovation but deplored its lack of disciplined scientific discourse—"being undisciplined, it all comes out like horseshit. But it's the very highest grade horseshit, like methane in the biological cycle." In passing the comment along, Marlin added that nothing could be richer than compost. When Macmillan demoted the book from a full-page and cover feature of its fall list to a half-page on the spring list, Bucky was happy to share the page with *Compost Gardening.*

Getting back to intertransformability, I had written a memorandum to the publisher describing *Synergetics* as "a book about models." When Bucky read that he agreed, but added

. . . humanly conceptual models, lucidly conceptual models, primitively simple models; and the primitively simple numbers uniquely and holistically identifying those models and their uniquely intertransformative number-value accounting.

Let us take the phrase "intertransformative number-value accounting" and try to explain it in plain English. If I were a teacher at the blackboard I might write "energy" and say that the finite physical Universe consists entirely of energy. The energy is in two forms: associative as matter or disassociative as radiation. Both forms are intertransformable, one with the other. I might go on to say that proton and neutron are intertransformable because if one transforms, the other does likewise, hence *inter*-transformative. Similarly, number-value accounting is just a way of measuring energy where the geometrical units and the numerical units are intertransformative. In synergetics, number is not an abstraction: each number has a geometrical identity as well as a numerical identity. The two are inter-transformative so that the number measurement of areas and volumes always comes out even, without fractions or odd numbers left over. No *pi*; nothing to the right of the decimal. Q.E.D., perhaps; but having said all this, the teacher is apt to be left standing at the blackboard long after all the students have gone out to play. Better let Fuller say it his own way.

5. Writing Out Loud

THE MAJOR CONTENTS of Fuller's other books derives from originally oral presentations. His best speeches are to young and sympathetic audiences willing to hear him through until the final thread of the argument is tied into place. This is a process that can go on for four to five hours; while the back rows have long since straggled to the parking lot, the young remain in a huddled trance around the podium.

Since communication is a two-way street, lecturing to others with "feedback-by-eyes" from the audience is his preferred mode of expression. He has an extraordinary ability, even in a large audience, to recognize the faces of friends he may not have seen for years, or to comment quite accurately afterwards when some of us in the front rows may have registered impatience or waning attention. Perhaps this intuitive sensitivity is partially a function of the deafness he developed in later life. Certainly he prefers feedback-by-eyes to questions from the audience. Sometimes when he is asked whether there will be a question period, he will answer, "Yes, after the lecture is over let us all go home and each of us ask ourselves our own good questions." Furthermore, questions—even in a small group—tend to distract from the broad geodesic structure of his argument. Or even if you put your question to him when there are just two of you, the response is apt to be Socratic. He just gives you the huge view of the largest imaginable patterns and then you have to go back and find the answer yourself.

His lectures assembled into a book—like *Utopia or*

64

Oblivion—have an immediacy of impact, but the
repetitions and overlappings are inevitable, though
Fuller regards this as a virtue rather than a deficiency. No one could ever have lectured with a
greater sense of obligation. Psychologically, he says,
he'd prefer to clam up, but his lecturing is the most
relentless form of self-discipline with every speech
demanding a "fresh inventorying of experience."
Direct expression to a responsive audience is in the
tradition of classic poets, the great teachers, and the
yogis. For Fuller, it affords an integrity of communication that he cannot achieve through any other
means.

Fuller's speeches are if not always impromptu, at
least spontaneous and extemporaneous in the
extreme. Even in the middle of a long lecture he will
not hesitate to pause and keep the audience waiting
for long self-conscious moments while he decides
which fork to take in the road ahead—he would say
that the audience participates equally in these
decisions of how the narrative should unfold. He
refuses even to cue television interviewers before
going on camera. He wants to preserve what he calls
the "pristinity" of the occasion; he will hear the first
question at the same time the viewers do. When
pressed for a subject—"What are you going to talk
to us about, Dr. Fuller?"—to put on the printed
program or poster, he will oblige with titles like:
Planetary Planning, Wood Design in a Dynamic
Technology, and Energy Resource Alternatives; but
the title is irrelevant except as a metaphor for the
larger synergetic theme. There is always only one
topic: the universe—whole systems. Journalist Barry
Farrell says that Fuller's talks are like his geodesic
domes: their structure does not reveal itself until the
last strut is in place. Just as the pattern of what he

writes is often not apparent until the last clauses complete the circle of argument. Just as he won't mail that letter until the last margin has been filled.

Of Fuller's first ten books, three were composed exclusively by himself and written out in longhand: these are *Nine Chains to the Moon* (1938), *No More Secondhand God* (1963), and *Intuition* (1972). Most of the other books are compilations of letters, lecture transcripts, and largely ephemeral—but never casual —forms of oral expression. His first full-blown literary work, *Nine Chains to the Moon,* is unlike any of his subsequent works in that it was conceived and executed as a single *book.* It contains the germs of all his later thought, and it remains today the most comprehensive—though unsystematic—statement of his philosophy; it expresses his program for a world economy as well as his vision of a psychological and philosophical kind of geometry. *Synergetics* embraces only the latter—the geometry —and thus falls short of wrapping up the entire Fuller canon. Except for footnotes and brief references departing from the main geometrical argument, *Synergetics* excludes his economics and industrial design programs, Spaceship Earth and the World Game themes. This was a deliberate strategy and virtually a practical necessity; but Fuller regards this as an artistic deficiency of the book and it is understandably a cause of distress to him today. His notebooks of the 1940s and 1950s reveal a grand pattern of presenting his geometry by starting with the most abstract conceptual patterns, which become more detailed as space-filling tensegrity structures merge almost imperceptibly into geodesic dome designs with all the detailed exposition of the patent applications, and ending up as engineering manuals for on-site assembly of the domes themselves. This

plan has never been realized, but there is no reason
it could not be, given the funds for an edition of half
a dozen volumes and the full production facilities of
a good university press.

In a review for the *Washington Post,* David Park,
a physics professor at Williams College, says,
"*Synergetics* is a book about structures, and so we
need not bother . . . about the rest of the ecological
and political thinking, though Fuller is all of a piece
in the way a more logical mind is not, and to divide
him is to damage." Park is saying "more logical" in
the sense of more conventional or more culturally
traditional; or maybe he isn't, maybe he thinks it is
illogical to combine structure and ecology. It is a
recurring dilemma: when editors organize his
material in what seems to them a logical way,
inevitably to Fuller it seems messy; when Fuller puts
it all back in what he calls a scientific sequence, it is
apt to seem messy to the editor.

For our working sessions in Washington I would
prepare our stacks of manuscript drafts and
drawings, not only in order of priority, but in what
struck me as a logical sequence. Bucky's own sorting
patterns were of quite a different order, and if he
didn't like the openers, he would not hesitate to deal
a new hand. He has a flair for reopening the narrative at some new place. And so he would reshuffle
the manuscript pages as he rearranged his trains of
of thought. After his departure, it might take me the
better part of a week to sift through the debris and
accommodate the new patterns he had imposed—to
the extent I could discern them—or, failing that, to
restore lamely the previous order as I had perceived
it. (I did not feel put upon; to me it would not be
nearly as much fun to try and edit the product of a
conventionally tidy mind.)

Prior to *Synergetics,* the best expressions of Fuller's epistemology, his geometry of thinking, are to be found in two mind-bending essays which he wrote out in longhand and which derive in no way from lecture transcripts or oral expression. The first is "Total Thinking"—he has a genius for arresting titles—written when he and Anne were at Black Mountain College near Asheville, North Carolina, in May 1949. The years at Black Mountain were among Fuller's most artistically productive periods —and no wonder! he was in the stimulating company of such bright lights as Josef Albers, Willem and Elaine de Kooning, Merce Cunningham, Ruth Asawa, Richard Lippold, Arthur Penn, and John Cage. "Total Thinking" emerged in print as chapter 12 of *Ideas and Integrities.*

The second essay is "Omnidirectional Halo," written ten years later (1959) during two rainy winter days when he was holed up alone with a cold in the Hotel Benson in Portland, Oregon. "Omnidirectional Halo" appears as the last two articles of *No More Secondhand God,* a collection of mostly verse items, marvelous in themselves but something of a dog's breakfast bound between two covers—quite possibly the result of having cleared off whatever happened to be on top of Fuller's desk at the time. It is characteristic of his disregard for conventional organization that his essay is broken in half, with the first 12 pages appearing as an introduction to the last 32 pages. If I had to explain *Synergetics* in one simple statement, I would say that its aim is to substitute "in, out, and around" reality for "up and down" thinking. "Omnidirectional Halo" describes how this can be.

These two essays are prodigies of Fuller's peculiar prose-poetry: dense, cryptic, difficult, provocative, uninhibited. He describes them as intuitive and fleeting glimpses of omniscience. They are not everyone's dish. I know that some philosophers— at least Bertrand Russell—tell us that the unexamined life is not worth living. Epistemology, abstract thought, philosophical speculation drive a lot of people up the wall. Intellect has nothing to do with it; it is a matter of taste or temperament. Many intelligent people are suspicious of mere logic: for them, the line between dialectic and sophistry is never clear; they are like the lady who had never done anything wrong—once she thought she had, but she found that she was mistaken. The briefest grounding in semantics can make any writing unreadable for some, until the self-conscious preoccupation with the thought processes wanes and the normal reflexes and filters become restored. Few people can read "Total Thinking" and "Omnidirectional Halo," but for those who can, they are the best epitomes we are likely to have of Fuller's philosophy.

The original plan for *Synergetics* as I inherited it was to incorporate "Omnidirectional Halo" verbatim and intact as a separate chapter in an anthological manner. I thought it would be retrograde to repeat what was already in print, and I knew that some of the concepts it advanced had been further refined in Fuller's thinking in the intervening years. He reluctantly entertained my assurances that all the items covered in the essay would be included in *Synergetics,* but stated afresh and in a larger context. This was done. All of the substance of these essays had been fully recapitulated in my voluminous index-card extracts—together with his related statements on these topics, both before and since. Thus the means were at hand to provide him with, in effect, new first drafts in his own words. If it could be made easy for him to review, reshuffle, and rewrite these extracts in the style of his philosophical essays, I felt sure we would be on the right track.

In my dog-eared copy of "Omnidirectional Halo," virtually every paragraph and many individual sentences are keyed to the section-number citations where they reappear as rewritten in *Synergetics*. One long sentence survives verbatim: I went over it quite a few times with Bucky, but he was totally satisfied with the way he had expressed himself and saw no way to improve it; certainly there was no reason to change it for arbitrary considerations. The result was not plagiarism, even self-plagiarism, but healthy reconfirmation.

By mid-1971, Fuller had started rewriting enough of my first drafts to assure me that the procedure would be effective. It was my early practice to cite, in the right-hand margin, the exact source and date of every phrase, clause, or sentence, since all I was doing was making a new, systematic, and sometimes exhaustive construct of the things he had said and written from a whole gamut of sources and in a variety of different contexts. Some very useful working techniques came out of our early sessions. I could see that the source citations were such a distraction to him that I had to stop putting them in the margin. Afterwards, I listed the citations on carbon copies, on the back of the manuscript pages, or in a meticulous card-file log.

Bucky usually accepted the authenticity of my source material, but there were times when he did not recognize his own language and little wonder, since some of the quotations stemmed from the 1930s. His occasional doubts caused intense anxiety, and I knew I would always have to come up with precise citations if I was to retain his confidence. On the few occasions when it took me 20 minutes to an hour to locate the precise source and context of a quoted sentence, the tension between us would mount; he would find it hard to concentrate, and I would be suspected of error or fabrication until I found satisfactory documentation.

I had bought heavy 14-inch mimeograph paper stock—like foolscap—to use for our drafts; it was of a distinct bright canary color to be used exclusively for work on the book. It became a signal between us, as he always sensed my impatience at losing precious time whenever the pages we were working on were not canary-colored—those competing and overlapping compositions again. At first I double-spaced the typescript, using wide margins for the inevitable graffiti.

These margins also enticed Bucky into the liveliest of geometrical sketching, spontaneous but meticulous drawings in pencil, ink, or felt-tip pen (or a combination of all three). Clusters of closest-packed spheres or great-circle spinning or four-petal tetrahedral flower buds would verge into the typescript to help him visualize the harmonic patterns he was working on. He cannot draw anything static; his figures all torque or twist or penetrate each other. No two are ever quite the same, relics of his accompanying verbal descriptions, interwoven into the manuscript and forgotten after they have been elucidated. It was like drawing out loud.

For Fuller no book—not even a single manuscript page—is a static takeout. He seems never to employ any art form without transforming the medium itself. (See illustration.)

Unlike Bucky, I never felt the lack of a typist or

(Modelability)

(<u>Pi</u> and Synergetic Constant)

982,63-65

Sphere = Vector Equilibrium in combined

four-dimensional orbit and axial

spin. Its 12 vertexes describing

~~four great-circles and six axis~~ *SIX* → SPINNING ONE AXIS

WITH ALL 25 CIRCLES/CIRCLES WHILE

Produces 2 *of A* ~~spin-profiling~~ *Superficially = perfect sphere*

VE HAS ALSO 25 GREAT CIRCLES (Sec 8) ---
OF WHICH THE TWELVE HAVE TWELVE X12
AND THERE IS " FOUR " FOUR AX11
Vector *THREE* *SIX*
Equilibrium = ~~Sphere~~ at equilibrious, ergo *THREE.*

25 Gt. Ci.

zero energized, ergo

unorbited and unspun *state.*

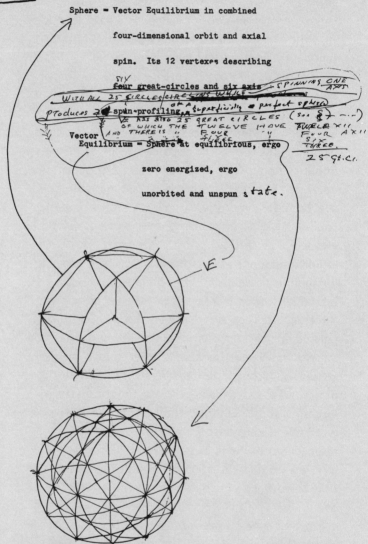

Fuller's drawings on manuscript of
Synergetics at secs. 982.63 to 982.65.

stenographer. Only by typing every page myself could I have the same kind of freedom to add, subtract, and rearrange as he had with his pen in the margin. Not everything he wrote in the margin reappeared in the retyped version; I was trying to police the duplications and would sometimes omit what he had written in, where I knew it was going to be treated in greater detail a few pages later on.

I would often beg Fuller to read a new draft section through first before picking up his pen to correct it, but I never succeeded; there is something in his visual-mental coordination that forces him to concentrate on the single page before him and prevents him from dry-running through four or five pages in momentarily suspended intervention. He is afraid the reader will function the same way, so he crams details and references in that first paragraph anyway. Who knows, he feels, maybe the reader will never get to that next page. Bucky and I approached the typescripts with completely different focal lengths.

Eventually, we had sessions where he would completely rewrite 20 to 30 manuscript pages in a day. I never felt he was drawing Dali moustaches on our Mona Lisas. The great reward for me in our working together was to find the fresh thoughts and phrases, however convoluted in the balloons and interlineations. I could never get him to turn the page over so he would have more room. Nor would he ever carry an insert over to the blank sheets of paper with which I would surround him. Sometimes he would crowd in six or seven lines to an inch. Whenever I asked him why he had to get it all on one side of the page, he would explain, "that's just the way I see it in my mind." There was a rationale:

RBF: You're not thinking if you know what you're going to be thinking about. If you know where you're going to come out when you start to cerebrate, then that's not thinking; that's merely brain-sorting.

This is why, when I am drafting in the margins of typescript redrafts, the whole process of thought-catching is so sensitive that I do not know how far this intuitive following of unfolding thought is going to lead. I do not know whether I need more space. I cannot interrupt to turn over on the back of the page or go over on to a new sheet.

The thinking is not linear, but it has to be expressed in a linear manner; it is a matter of recording an unexpected omnidirectional involvement in a linear writing or graphing pattern.

In our first session in Sarasota, I learned that I was not leaving enough room for him to add his write-ins. I hadn't fully learned the lesson from the Carbondale typist. So I started making the typescripts triple-spaced and the right margins two-and-a-half to three inches blank. In fact, the crazier the text or dictation seemed to me—he would continue to dictate, not literal text but just thinking out loud— or the more unsure I was of the material I had assembled, the broader would I leave the right-hand margins as a special invitation to rewrite. There was nothing surreptitious about this; I tried to tell him what I was doing. I don't know whether he ever caught on to the scheme; sometimes my three-inch-margin pleas to rewrite would come back virgin, other times they would make me wonder whether I was collaborating with Jackson Pollock—or a monk replaiting the Book of Kells.

As an example of the latter, let me give a before-and-after version of the text with an illustration of what he did to rewrite it. (See illustration.) I had found an intriguing paragraph from an unpublished preface to a book written by his Oxford friend,

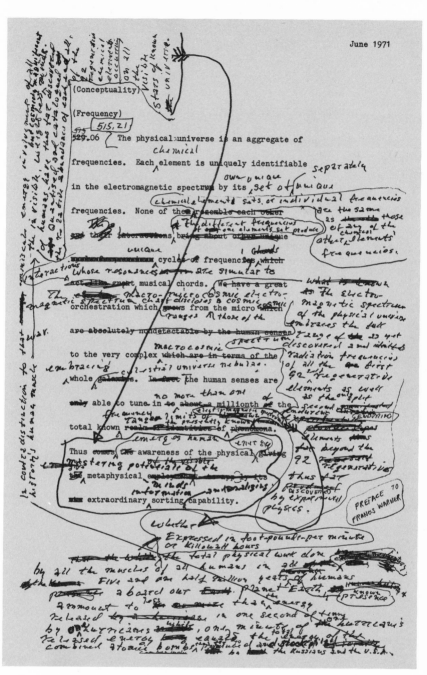

Fuller's manuscript revision of Sec. 515.21 of
Synergetics, June 1971.

Francis Warner. As often happened, Fuller's preface had missed the publisher's deadline, so the piece was in a sort of literary limbo. I thought one paragraph would fit very well into his discussion of the phenomenon *frequency* at Sec. 515. So I put the appropriate running heads at the top of the first draft and left a right margin of almost three inches, with results as follows:

Before:
The physical Universe is an aggregate of frequencies. Each element is uniquely identifiable in the electromagnetic spectrum by its frequencies. None of them resemble each other and their interactions bring about other unique cycles of frequencies which act like great musical chords. We have a great orchestration which grows from the micro which are absolutely non-detectable by the human senses to the very complex which are in terms of the whole galaxies. In fact the human senses are only able to tune in to about a millionth of the total known realm of identities of phenomena. Thus comes the awareness of the physical giving the metaphysical employment—to apply its extraordinary sorting capability.

After:

The physical Universe is an aggregate of frequencies.
Each chemical element is uniquely identifiable in the
electromagnetic spectrum by its own unique set of
separately unique frequencies. None of the chemical-
element sets or individual frequencies are the same as
those of any of the other chemical elements' frequencies. The different frequencies of one element's
set produce unique cyclic-frequency interactions whose
resonances are similar to musical chords. The
electromagnetic spectrum of physical Universe embraces
the full spectrum range of as yet discovered and
identified radiation frequencies of all the first 92
self-regenerative, as well as the only split-second
enduring elements beyond the 92 self-regenerative thus
far discovered by experimental physics. The macro/
micro-cosmic electromagnetic spectrum chart discloses a
cosmic orchestration which ranges from those of
the microcosmic to the very complex macrocosmic-
embracing whole celestial Universe nebulae. The human
senses are able to tune in no more than one-millionth
of the total known frequency range limits of the presently
known electromagnetic spectrum. Whether expressed
in foot-pounds-per-minute or kilowatt-hours, the
total physical work done by all the muscles of all humans
in the five and one-half million years of known
presence of humans aboard our Planet Earth, amounts to
less than the energy released in one second of time by one
hurricane, while one minute of one hurricane's released
energy equals the total energy of the combined atomic
bombs thus far produced and stockpiled by the
Russians and the U.S.A. In contradistinction to this
minuscule energy involvement of all history's human
muscle, the invisible, weightless, but cosmically
magnificent minds of humans have thus far discovered,
quantized, and catalogued the relative abundance of
each and all of the 92 regenerative chemical elements
occurring on all the visible stars of known Universe.
Thus emerges human awareness of the physical-energy-
mastering potential of the metaphysical mind's
extraordinary information-sorting and -analyzing
capability.

There is nothing in "After" that was not implicit to Fuller in "Before": in the rewrite he has just permitted the reader to share with him a little more of the philosophical foray. He must have been satisfied with the result as the only change he made on the galley was to reduce the human life-span on earth from five and one-half to two and one-half million years. The final version may be found at Sec. 515.21, the result of only one rewrite.

One of the advantages of writing over thinking out loud is the availability of the shouts of capital letters and the murmurs of lower case. Fuller had sprinkled the text of *Nine Chains to the Moon* with words and even whole phrases all in capitals, and he suggested reviving the practice for our book. I demurred because I found the big-blocked words distracting to the eye and because the reader prefers a nudge to an elbow in the rib. So some consistent house rules emerged in the process of composing *Synergetics*. Fuller likes initial capitals, but in a way quite distinct from William Blake. For Fuller, nature —like integrity and principle—is abstract and generalized and thus never capitalized. Universe is a place like Pleasantville or Baltimore and deserves a capital letter. Universe is physical and the physical is always special case, i.e., realized in space and time, and thus capitalized. You cannot have a generalized boat, or canoe, or schooner, not in physical reality. Realized physical boats have names. The Universe—ours—has its name.

The same with Earth; when we dig up "earth" with a spade instead of "Earth," it is because we have failed to recognize the unique identity of the stuff of the third planet from the Sun. (*Sic.*) I once argued with Bucky that at least moons were generic;

but not ours, there is only one and it's the Moon.
Conversely, God, in *Synergetics,* is god because all
the associations of God are irreparably anthro-
pomorphic. God, he explains, is part of the thinking
process of every man.

Fuller's self-disciplined precision of speech and
his close attention to Heisenberg's indeterminism
prompt him to unfamiliar uses of the word
approximately. He says "approximately everywhere"
and "approximately none" and even "approximately
one." He says he uses the phrase "approximately
everywhere" to make the everywhereness coincide
with the modular frequency characteristic of any set
of random multiplicity.* This practice—as well as
our peculiar capitalization—caused difficulties with
Ken Shaw, our erudite copy editor at Macmillan.
Shaw would say you can't say "approximately one."
And I would answer no I can't, and you can't, but
he can.

The way we organized our work we had agreed
to define our terms precisely when they were first
introduced, not only Fuller's invented words—57 at
last count—but conventional terms he employs in a
special way. Each new topic was captioned in the
chapter outlines and there was an overlapping of
redefinitions and cross-references. This led us to
think we might dispense with an index. We felt the
built-in repetition might overload and degrade the
page references to specific topics. Moreover, an index
is inherently antisynergetic, encouraging the search
for nonexistent "building blocks," the erroneous
pursuit of the whole from the particular. Our
rationale for this may have been a little contrived.
We recognized that—ultimately—an index to the

---

*RBF letter to Dr. Urmston, 8 October 1964.

complete work would be an essential tool for the reader. With all the final rush of crowding last-minute text into production, any notion of an index had to be abandoned as a practical necessity.

In our spontaneous and impromptu, almost aleatory, capturing of fleeting thoughts into first drafts, we would often start off a passage in one context and then see it develop with successive redrafts along lines more appropriate to another section—or another chapter—sometimes inadvertently filling in a missing gap. I called these orphan compositions "floating inserts" until they were tied down with an assigned section number in the outline. On occasion, Bucky would suggest a locus that would never have occurred to me but which enhanced the implications of what had been going on in his mind. I tried to raise the question of where these floating inserts should be placed on many occasions; sometimes he would have a suggestion, other times he was oblivious. In principle, he applauded the virtues of my topical, section-number breakdown, particularly for ease of cross-reference. But my relentless naming of parts tended to obscure his focus on the whole. (After all, what were all those Köchel numbers to Mozart?) Though he had every opportunity to control or determine the broad outlines of our chapter organization, it was not an omnidirectional procedure, and he could never embrace it as his own.

I have never wanted to write a biography of Buckminster Fuller, not that it wouldn't be a great subject: the hero as thinker and myth-maker. What really interests me is not so much his program, but his mind, particularly those ideas he had never fully articulated. His whole life is a metaphor, an organic

artifact for the instruction of others. Mrs. Athena
V. Lord of Albany, New York, is writing an account
of his youth and growing up as a book for children;
in their first interview, Fuller suggested that she
mythologize his childhood. To document this kind
of a living myth would require much more
detachment and distance than I could muster. And
yet I feel a sense of keen privilege as one human
being able to devote his time to another to an extent
that must be rare among contemporaries. Most
scholars trying to enter the mind of another in any
prolonged and systematic way must deal perforce
with subjects—artists, writers, statesmen, thinkers
—long since departed. When I think of all the
professors of English or history devoting the major
part of their careers to studying a single figure from
the past—how they would envy me: I can ring up
the subject of my studies on the telephone any time
. . . to clarify a point, to mine his memory, to seek
a judgment, confirm a nuance . . . or save my
questions till we meet for two days every month
or so.

6. Buckminster Abbey

FULLER WAS IN continuously orbiting world travel while I was sedentary and rooted at my home in Washington. The pattern of summer schedules left us very little time for working together and one summer went something like this:

June: Honorary degrees from several universities or colleges where he would lunch amiably with the trustees, who paid no attention to the subversive content of his remarks; his real allegiance was with the students, and they were his only target. Bucky has described himself as an immigrant in the society of the young.

July: First to Gian Carlo Menotti's festival in Spoleto for a last embrace with Ezra Pound, whom he passionately defended for sharing his view of the irrelevance of politics to poetry—a point on which Pound had been a slow learner. Then Anne and he would be guests of Constantinos Doxiadis for a cruise around the Aegean ending in a conference on Delos. Bucky's lectures were a gentle rebuke to "Dinos" Doxiadis, the founder of *Ekistics,* the science of human settlements, for the same fallacy as that of Lewis Mumford: regarding the city as a proper object of study when the minimum considerable social pattern is the totality of Planet Earth. Only the whole big system works.

August: Theoretically, this month was reserved for the family's vacation on Bear Island in Maine, but it was interrupted, as usual, by a United Nations conference in Southeast Asia—what Fuller calls Austronesia—and by board meetings on the West Coast.

September: By this time the lecture circuit would be resumed with occasional side-trips for a college gaudy in Oxford or congressional hearings in Washington. Fuller told a panel of six senators that politics is obsolete: Who needs leaders?

As a reluctant vagabond—my children had once predicted, accurately, that I would be the only father in Disneyland with a tie on—I made various attempts to co-orbit with Fuller for brief periods and, except for a debacle at Bear Island, they invariably produced original material for the book. Bucky can go to sleep and wake up instantly, refreshed after only 15 minutes. He has tremendous powers of concentration and the ability to screen out the irrelevant, so even after weeks of separation we could pick up the threads of our writing—and even at odd hours or on days with an already crowded schedule. I couldn't take my typewriter into taxis and he didn't like me taking notes in the dining room of the Somerset Club in Boston, but we managed to waste very little time.

We had one magnificent uninterrupted working session in the top floor of the Blackstone Hotel overlooking the lake in Chicago on a Labor Day weekend when we were almost the only guests in the hotel. Uninterrupted and alone—except for Don Moore, Bill Marlin, and Brendan O'Regan who took part in most of our working sessions. Fuller was reminding us that energy has shape and that thought has shape and that these physical and metaphysical shapes manifest themselves as tetrahedra, specifically his discovered asymmetrical tetrahedra which he had long referred to as the A and B Particles. But the word *particle* stuck in his craw; it was not right because there are no things, no basic building blocks

—there are only relationships between events. At the Blackstone, we buried the particles and rechristened them the A and B Quanta Modules as they appear in the book.

Fuller asserts that his synergetic geometry integrates the landmark topological formulations of Leonhard Euler (the Swiss mathematician, 1707–1783) and the famous "phase rule" of the American physical chemist Josiah Willard Gibbs. Fuller had long promised me a sketch or an outline of how he combined, and partially transformed, Euler's visual analysis and Gibbs' energetic freedoms. He finally sketched it all out between flights on one of those glossy breakfast menus in the old coffee shop at Logan Airport in Boston. I sprayed fixative on the menu as soon as I got it home, but the smudges had already made it difficult to transcribe and some of the nuances may have been lost where it appears at Sec. 1054.20.

We had a wonderful week with Anne and Bucky at Marcel Breuer's new student union building at the University of Massachusetts in Amherst. They had just gotten in from London and TWA had misplaced Anne's bag; she just cheerfully bought a new toothbrush and remained serene until the bag arrived two days later. Every night, Bucky lectured alternately with the Maharishi Mahesh Yogi, whose guests we were in a symposium sponsored by the Students International Meditation Society. By day, Bucky and I were trying to write *Synergetics* despite interruptions from a constant stream of visitors, well-wishers, and just plain sightseers. One morning, Bucky was in unusually deep concentration, his gaze turned inward on some remote horizon of thought; he was just on the point of resolving some apparent contradiction in our text and his face was

like a mask of a Dalai Lama. Fuller was in as close
to a trance as I have ever seen him when a well-
meaning friend burst in on us to urge him to attend
a lecture by a young mathematician—in a classroom
seven floors below. Although he wakes from sleep
in an instant, this time it took us almost half a
minute to bring his consciousness back to our
everyday world so that we could explain the simple
invitation. He politely declined, fixed himself a fresh
pot of tea, brushed those particular pages aside,
and never regained whatever vision of Shangri La
had so entranced him. I felt the loss keenly as that
particular cosmic fish got away forever.

For a long time, Fuller had circling in his head an
elaborate metaphor for his omnidirectional system
of spherical coordination. His model described a
transparent plastic 20-foot miniature Earth globe
inside a transparent plastic crocodile inside a
transparent plastic whale inside a 200-foot
transparent plastic celestial sphere. He would have
preferred for us to build an actual full-scale model
but we had to settle for landing it all on paper
(Sec. 1110.00) during a week visiting the Hugh
Kenners at the Pepper Tree Inn in Santa Barbara.
Other such fish were caught in Aspen, in Carbondale,
in Fairfield, and in Bryn Mawr.

In lecturing, in writing, and in conversation, Fuller
often pauses to take stock, to review his tactics and
strategies, and reconfirm his objectives. This is one
of his many forms of self-discipline which to him
is akin to the navigator's taking fixes on his charts in
river piloting. One evening in February 1973, when
we were working together in the motel in Santa
Barbara, he handed me a sheet of paper in which he
had written the following description of our goals
and modus operandi:

Sonny Applewhite and I meet deliberately and premeditatedly, and thereafter find ourselves spontaneously, inadvertently hauling-in word-netted shoals (schools) of cosmic fish, i.e., epistemological pisces.

Sonny handles the ship, opens the holds, and heads our catch for the commonwealth harbors of humanity, while my task is to cast the nets of prescient apprehension in discrete directions in the omnidirectional ocean of Universe to be hauled in only upon unpremeditated observational embracements of ever-more-stabilely-generalized systems of ever greater and more incisive comprehension with which we may classify and sort our cosmic-fish catch of ever-multiplying Universe's special case experiences.

Our most regular base of literary operations, second only to Washington, was the Beverly Hotel on Lexington Avenue across from the Waldorf-Astoria in New York. It had the virtue of semi-residential suites so Bucky could make his own tea in the pantry and I could type in the sitting room while he napped. The switchboard and front staff treated us as old friends and it was a congenial place. I did not want to stop writing for meals but Bucky clearly preferred to. Fuller's staple, or at least his preferred diet at every meal, is steak, a diet that helps him to keep his weight trim, suppress a tendency to diverticulitis, and above all refuel those enormous stores of energy. His rate of metabolism would be the envy of a shrew. When young vegetarians express dismay at his beef-eating vice— it does not conform to their idea of his persona—he has to explain that the cows are eating much more vegetation and converting it to protein than he could possibly cope with at first hand.

Conveniently across the street from the Beverly was the Royal Scots Grill (staffed by Irish, Koreans, and a beautiful Israeli sabra) where steaks were available for breakfast. This became Fuller's club. We would put off visitors to mealtime when we would repair across the street to breakfast, lunch, or dine with a stream of publishers, dome builders, artists, students, and mendicants—Clark Eichelberger, Bill Whitehead, John and Magda McHale, Cyril, Joe Duffy, David Cort, or Peter Max. The Royal Scots Grill (since renamed the Belmont Stakes) became the Mermaid Tavern de nos jours. They were very good about turning down the Muzak.

As the book took shape it acquired a sort of critical mass attracting Bucky to spend more and more time in writing sessions at my home in Washington. Our longest session was 10 days. We live in a 100-year-old, pseudo-classical, white frame house on a high acre of land edged by ancient oaks. Of our four children, three were then at home, ranging in ages from 12 to 20. It was just as well that the house had abundant waste space so we could come and go without treading on toes. I made an office out of the large sunporch overlooking the grass to the west.

There were filing cabinets for the manuscripts, letters, and unpublished papers, as well as each meticulously numbered and dated sheet of the myriad of canary yellow earlier drafts—superseded but not abandoned. There were two bookcases with books only by or relating to Fuller and his geometry. There were two trunks sent from Carbondale containing the original files of Fuller's synergetics notes, notebooks, drawings, blueprints, and collateral clippings. And there were my twelve file boxes of 5 x 8-inch alphabetized excerpts of everything he had said or written about everything that I could get

A SYNERGETICS DICTIONARY

Being a Systematic, Nonlinear, Variorum Selection
and Self-branching Arrangement of the Concepts, Definitions
and Generalized Principles Discovered or Formulated by

R. BUCKMINSTER FULLER

Embracing the Whole Body of his Writings, Letters,
Lectures, Table Talk, and Marginalia on the Subject of
Synergetic Geometry and Philosophy presented
exclusively in the Author's own Words
as an Index and Adjunct to
His major Work:

SYNERGETICS

Expressed in Model, Sequence, Metaphor, and Aphorism as
Recorded, Compiled, Arranged, and Cross Referenced
for easy access by

E.J. APPLEWHITE

"Catalog of a Mind" is the subtitle for the author's
22,000 extracts and index to Buckminster Fuller's
geometry and philosophy. They were the basic
resource for *Synergetics* and have been reduced to
microfiche for the use of Fuller and his students.

my hands on. At one end was my desk and my old Underwood standard typewriter surrounded by low tables on which the file boxes were arrayed like the manual of an organ in such a way that I could reach any of the 22,000 excerpts from where I sat while typing. (See the title sheet for these files on opposite page.)

At the other end of the porch was a large round work table (once a family dining-room table, now telescoped without the leaves) with a lamp in the middle for Bucky to write at. We also had sticks, dowels, cardboard, paper clips, and rubber bands for impromptu model making. I first called our office-porch "The Buckery," but this had a connotation displeasing to Fuller, so we changed the name to "Buckminster Abbey."

Coming in the front door or coming from upstairs you could see Bucky through the glass doors from the living room, sitting ramrod straight at the round table on the porch. He would be drawing or writing furiously or staring out at the oak trees. The family knew that they would not bother him going back and forth to the kitchen because he was always so deep in thought. Maybe his deafness was an asset. And when the children did interrupt he was always far more tolerant than I. Fuller has a genius for putting interruptions and the unexpected to good use. What for me was distracting was for him happily synchronous. I was collaborating with a man infatuated with problems, who was looking for some meaning in every interruption: confident, as he says, that the almighty makes no mistakes.

On Bucky's round table there was a hot plate for his never-empty pot of tea—the second staple of his diet—which he consumes like bunker fuel around the clock at a level just below toxicity.

RBF: You know how I have so often described the thinking process as a dismissal of irrelevancies . . . a parting of the strands. But there is also an element of conscious input to thinking. Certain thoughts have to be held in suspense in momentary irrelevancy like the airport holding patterns of planes coming in for a landing.

Something like this happens when I am concentrating so hard that I put off going to the bathroom. Anne often says to me, Bucky, for heaven's sake why don't you go to the bathroom—you've been jumping up and down for half an hour. . . . And I do. . . . I just keep postponing the input of the tea and the output of the pee.

Science and the physical world give you deliberate stresses with the results of forcing you to do your best work. It's like rushing to catch a train. . . . Or the way a drowning man sees all the events in his life going by. . . . I get so preoccupied with my thoughts that I cannot find where my fly is and I make a mess in the bathroom. . . . When the rest of life is pushing in on me it forces you to make a comprehensive review . . . instead of waiting for your life to go by while you're jumping out of the building. I just build up so much overload and that's how I do my best thinking.

He abandoned alcohol as a drink in the 1940s, and his later work shows a strain of sobriety of which *Nine Chains to the Moon* is happily devoid. There are passages of *Synergetics* which in cold print suggest the evidence of too much tea—although he is usually careful to guard against overstimulation by switching temporarily to grapefruit juice. Even though Bucky is personally immaculate, after two sweltering hot nights we found his pillowcases had pink stains that could be diagnosed only as tannic acid having seeped through his scalp.

Except for the total unpredictability of his schedule, Bucky was always the ideal houseguest.

His arrival would sometimes be heralded by an unexpected shipment of aluminum rods or wooden dowels or plastic struts. (In the days before his elaborate typed itineraries, we always signed for any odd shipments without question, knowing that we could expect the model-maker to follow within a few days.)

Fuller is not nocturnal by nature—as the Hewletts were; they could not retire before midnight without a sense of defeat—but he is often unable to stop working until he drops from exhaustion. I lack this kind of compulsion and though I can type quite happily from 8:30 in the morning to 6:00 in the evening seven days a week, I generally prefer not to work after dinner. He, on the other hand, would often work till 2:00 in the morning and—whether she was with us or not—I would sense and share Anne's strong disapproval. On two occasions when my wife, June, got down to start breakfast, she found Bucky still scrawling away furiously, having moved to the dining table where it was warmer. He had been oblivious that dawn had come, but he would instantly sense the domestic imperatives, pick up his papers hastily, and whisper, "I was just going to bed, darling."

When his meals did not coincide with ours, which was often, he would fix his own steak, fruit, and Jell-O. Being a well-trained sailor, he always cleaned up the pans in the galley. He takes a childlike thrill in finding and using new kitchen gadgets and was always bringing in new electric percolators or manual can openers. And he loves to sharpen knives; they were always in great shape after his visits and June would bleed for days.

With June and the children, Bucky was never less than gallant, always full of love—often verging on the fatuous. He is an enthusiast of family life and he liked to plunge into all the trivia and tensions, the by-plays and arcana of teen-age communication; he reveled in all the inside jokes and insults, whether he could follow them all or not. Bucky and my family found each other sometimes incomprehensible yet always fascinating. He clearly loved to share in the family life and realized he was totally accepted as a member by all of us—so much so that the kids always felt free to question and tease him and otherwise treat him as yet another relative. They had enormous respect for him, but no awe.

When friends dropped in I would want us to keep on working in the office-porch even though they could see us through the glass doors. Bucky had much better manners and he would rather talk than write anyway. In fact, he always loves a party and I was stifling his gregarious instincts with my rude determination to get on with the book.

We had everything we needed: steaks, tea, tranquility, and, what was very important for one who worked with Fuller's intensity, the opportunity for him to peel off upstairs for a nap whenever he would begin to nod after uninterrupted hours of work. Bucky's costume was usually as informal as our schedule. He once had a wonderful suit made in Montreal with the vest and trousers all of one piece, but of course you wouldn't notice this until he took his coat off. More often he would sport a long, thin, bright-green dhoti-shirt from New Delhi, but his favorite was a loose Japanese *yukata* and fiber slippers from Tokyo. During winter storms the porch was often icy, but he would never put on a sweater unless June told him he had to. His energy kept him warm. He was always neat, immaculate, very tidy in dress and work habits. Our working sessions were times of pure pleasure, even high spirits and hilarity, with unfettered communication between us. We interrupted each other all the time with even second- and third-degree interruptions interrupting the interruptions, but we were always able to get back to where we left off.

Whether we were working together at 3200 Idaho Avenue, N.W., in Washington or meeting in a motel in a distant city, I always had plenty of time to organize our various concurrent drafts in an order of priority that determined our working agenda. We seldom finished half of what there was to work on, and so we had an invariable rule: always put the most crucial drafts on top of the pile. We would discuss this explicitly; if we had only two days of work before Bucky was going to be hit by a bus, what would we most have wished to put his energies on? What was important, not in terms of any arbitrary schedule of chapters for a book, but what was philosophically most important in terms of an original breakthrough or a critical definition. Bucky would get so excited when he got a really new idea: that was when he first started to call them cosmic fish. In my copy of *Synergetics* he was later to write:

To: Sonny, my cosmic fishing partner who always lets me sit down aft in the fishing seat to have all the fun of catching and playing the fish while he handles the boat, standing at the wheel and engine controls, letting me think he is enjoying it all as much as I do!

We wrote in the constant company of posterity. While our thoughts were not morbid, they were preoccupied with intimations of mortality. (Five years ago he said to me, "I'm 76, I always assume I've only got a few more minutes.") Even if we only had a half hour between planes at the airport,

93

we always dealt first with the papers on top of the pile in the briefcase. For Bucky, this tactic was a guarantee against trivializing. Since he could spend only a small fraction of his time writing *Synergetics,* it was all the more imperative that he work at all times on the highest-grade ore and at the cutting edge. There was no time to indulge his pedagogical reflex of starting every session from scratch, recapitulating the whole context in his instinctive synergetic way. For me, this was the one way I could exercise some initiative; it was an effective way of focusing his attention on the unresolved problems— or at least as they so seemed to me.

Putting the important papers on the top was only part of it. We usually had three piles on the round table in the sunporch: (1) the most important (canary paper), (2) neglected drafts but not so critical (canary paper), and (3) other of his work in progress not related to *Synergetics* (white paper). Very rarely some kind of cul-de-sac or mental block would intervene and it would only make sense to turn to something else, some easier piece of descriptive writing. The third pile of white typescripts unrelated to the book provoked a distracting tension between us. He knew that if he spent too much time on the third pile I would begin to get restless; I knew that if he was preoccupied with some current composition it was hopeless to expect him to concentrate on the book. Sometimes he would work on all three piles in rotation, in which case themes and threads of arguments from one pile would find themselves being woven into drafts on the other piles; sometimes the effect was synergetic, other times it was merely fortuitous.

My greatest thrill, the keenest satisfaction in working with Fuller, was coming down early in the

morning and finding a stack of manuscript pages he
had rewritten way into the night before. He would
leave a note with the date and the exact time he had
knocked off: "Tout finis which rhymes with 2:15
A.M., and that's that." Or "Safe landing after half-
century nonstop flight 3:10 A.M." Or "These pages
are rich and exciting; we must bring them to
publication at the earliest possiible moment." And
sometimes hints of dark foreboding: "I am feeling
myself being physically precessed into a strange
ephemeral condition. It may be I just need a little
sleep. Only God knows. I don't know." He loved to
work from cleaned-up drafts and would politely
implore me, "Because this is the most controversial
of my work . . . it should be very neatly presented."

So I would get up early and usually have two or
three hours to work on cleaning up the drafts on
the typewriter before he got up, ate his breakfast
steak, and was ready to start on the next day's
sessions. For me, this was a highly creative business
and far from routine. I never retyped except with
concentration on substance. Here was the opportunity
to suggest combining certain themes, checking
assertions against previous statements, picking up
cross-references, depending heavily on my 5 x 8
file cards. The first two or three drafts of some new
topic might come out of the blue (at least out of the
blue to me, but probably not to Fuller who always
sees events in a more elaborate context than he has
time to articulate), and I would have to handle them
as floating inserts. As the pattern of thought or
geometry developed, I would see where they might
fit into one of our chapters under construction. All
of our manuscript pages had two running heads, one
for the chapter topic which was to go on the verso
pages and another for the chapter subsections to

be printed as running heads on the recto pages.
Gradually, I would present these more finished drafts
to Fuller in the context of our agreed chapter
outlines. He was ambivalent, sometimes enthusiastic,
sometimes indifferent about questions of organization.
Our procedure was very flexible. We could and
often did make enormous changes at the last minute.

Though it may not appear so, our modus operandi
put a high premium on spontaneity. As any one who
has ever dictated or rewritten a page knows, there
is a terrific compulsion to have the clean copy
returned almost instantly. In periods of intensity
we would work with only one paragraph to a page
and go through two or three generations of drafts
in an hour, working at least several pages at a time
as paragraphs would balloon omnidirectionally
answering their internal imperatives and without
regard to their ultimate sequence. I would be typing
rapidly to get a new paragraph back to him when
some completely new thought would come to him.
Since my typewriter was engaged, I would grab a
yellow legal pad to write down what he was saying.
Then I might read back in incredulity what I thought
he had just said. But by this time a third new
thought had come to him, and he would pace with
excitement, sipping tea while a new stream of
words—with very little syntax and sometimes no
predicates—would come bursting forth. To our
great delight and excitement we would find ourselves
in a third degree of interruption, with each
succession providing an implicit context for the
previous ones. While I was trying to field and sort
these out in feverish activity, he could not refrain
from coming in with a fourth or fifth degree of
interruption. It might be an hour until he returned
to the table and got engrossed again in the

manuscript; then I would have time to sort and
type up these various gestalt bursts and—with my
compulsion to name and label things—see whether
they were various aspects of the same topic or
merely concurrent and unrelated. It was an intense
and highly satisfactory form of communication:
mostly oral for him and, except for my constant
questions, mostly written for me. Confronting Fuller
with Bucky, confronting self with self.

Fuller likes to use tape recorders. Not only are
almost all of his lectures recorded, but often informal
sessions at home or office—and even at mealtimes.
This stems partly from his constant concern that
none of his cosmic fish—not the least minnow—
should slip through the net of his discourse. The
process is also an aural counterpart of his
documentary chronofiles. People ask him as a
perfunctory courtesy if he agrees to being tape-
recorded, and he will answer, "Of course, it can all
be on tape. I don't have any off-the-record life."*

As far as I know, Fuller was the first to use the
word chronofile. I remember its use from when I
was in the Navy, but Bucky says he coined it before
he was in the Navy.

Fuller's decision that his whole life should be on
the record was made at the age of nine. His private
compact made at that early age became a lifelong
chronofile project, an unrelenting and exhaustive
exercise in self-documentation. He was determined
to save every letter, paper, report card, bill, traffic
summons, love letter, petition, IOU, or collection
notice—favorable or unfavorable—reflecting on
paper the total record of one individual and his
relations with family, friends, society, and institutions

*RBF to Wm. M. Wolf, Philadelphia, Pa.; 15 June 1974.

at large. As Fuller says, "I'm the only guinea pig I
have"; and he wants to share its record with
posterity without any selectivity, discrimination, or
censorship. His secretaries seal up all the original
letters and clippings in bound boxes every month,
having forwarded him Xerox copies from which he
writes the replies, often right on the face of the
copies. Carbons of the replies also go into the
chronofile and, if they relate to the geometry, with
copies to me for processing in my index cards, which
I have subtitled the "Catalog of a Mind."

In addition to the letters, hundreds of cassette
tapes would be amassed in the course of a year, and
I got some very useful material for *Synergetics* from
this electronic resource. The worst part of it was
that very few of the recordings have been
transcribed and those that are transcribed tend to
be of very poor fidelity unless done by the rare
person familiar with his ideas and way of talking.
Of those who are familiar, few have the patience.
(One hard-pressed stenographer had once typed
"fascia birth"—whatever that meant—for what
could only have been "spaceship earth.") Thus I got
material for the book, but most often at the cost of
having to transcribe the tapes myself—which is how
I have spent many days of tedium.

As many times as I have heard Fuller speak, I
have yet to come away from a talk without having
heard something new, either new notions or new
phrasings, or both. And so it was with the tapes; I
would have to sit listening through three or more
hours of tapes to glean as little as one or two new
paragraphs or a couple of sentences of fresh grist
for my card files. And the tapes were another reason
why I felt that Bucky and I were working together
in Buckminster Abbey every day—no matter what

part of the earth he might be touching base at in
his actual itinerary.

Bucky would have liked to use tape recorders in
our sessions together, but I always avoided it. Just
the subconscious registration of the tape reels
methodically turning enticed him with the prospect
of an ultimate audience of posterity. When he had
such a constituency in mind it was difficult to
compete for his attention. Recorders also tempted
him to talk not faster than he could think but faster
than he could articulate. When he was watching me
write it all down as he spoke, I could glance up
with a glimmer of recognition or a look of skepticism
or profound doubt or even dismay, and he would
then refine and clarify on the spot. I could read back
what he had said, and together we would work out
the syntax. (Fuller has contempt for syntax and
grammar as such. I am sure he sometimes regarded
my pedestrian insistence on predicates as tedious and
irrelevant—but in the end the result was synergetic.)
It was frustrating for him to have to slow down,
but with the instant readback of my makeshift but
live dictation I could confront him with himself in
the most intimate and effective way.

For similar reasons I never wanted a stenographer
or secretary. I preferred the enforced slower speed
of my own super-hunt-and-peck typing, with which I
could improve, clarify, refine, and cross-refer, as I
worked through each redraft. Over the past five
years, I have typed an average of five or more
redrafts of an entire 800-page book, and my typing
speed and accuracy have benefited accordingly.

Our Buckminster Abbey sunporch became an
unfailing invitation to happy days of working
together. Our quarrels and disputes were rare. One
of us could sense a confrontation coming and head

it off by dropping the matter or by conceding a
point before tempers flared. Bucky was always
extraordinarily sensitive to my mood—as he was to
everyone else's—and I could never be annoyed or
frustrated for long and pretend to be otherwise or
conceal the fact from him. He might on occasion
ignore my displeasure, but he was never oblivious or
indifferent. The major battles we had were over
what I have come to call his "expository reflex
syndrome."

Fuller deplores reflex thinking and reflex
expression. For him, reflex is physical—like a
chicken with its head cut off—but it is not life. (This
is central to his rationale for abortion: that the
essence of life is awareness, not conception or
physical increase. Life is the conscious part of us,
not the physical part of us.) He dismisses reflexes,
yet I have found over the years of knowing him that
there are about a dozen topics or themes, or deeply
buried complexes, that trigger within him a profound
and urgent compulsion to retell and embroider long
and elaborate parables, like the Ancient Mariner to
the Wedding Guest, like the Polynesian navigator
with his chants, like an ancient litany from lost
books. He gets so wrapped up in the retelling that
he cannot escape and has to act out the whole
sequence in dramatic detail.

RBF: *What you call my expository reflex syndrome
is like pole-vaulting. You have to speed over the
approach path again and again before you clear a new
height of the bar. Then new heights provide new
viewpoints and their new vistas.*

EJA: *Yes, I respect the achievements of your tactic—
but I find the method tedious and frustrating
in practice.*

RBF: Yours is the trying, vicarious task of the track coach wearily watching the pole vaulter's day-after-day repetitive practicing. Vaulting records are only rarely surpassed.

You are a generalist, Sonny, but you tend to start from the particular. When you go to the library you could start out in any row of stacks, or you can go to a file and start at any entry and work toward a general pattern. On the other hand, I am not only a generalist, but a comprehensivist; the difference is that I am a design scientist. I am interested in the question of how did we get at this particular point? All the time I want to tell the reader how we arrive at the thoughts we arrive at. You don't have this design-scientist compulsion to describe every step of the way.

An example of this syndrome is his metaphor of "the deliberately nonstraight line," describing an endlessly regressing rope or line hung on nails or pegged in the sand, halved, quartered, and eighthed into ever higher frequency ad infinitum until it becomes a microscopically wavy line. It is a crucial part of his geometry; the hypotenuse of Pythagoras now in the hands of Fuller. He takes an enormous poetic delight in having his listeners share the graphic transformations, the remarkable properties of this simple rope and the kind of line it describes (Sec. 522).

The real snag was that we were about to wrap up the first third of the manuscript to send to the publisher and "The Deliberately Nonstraight Line" was one section of the book above all that had been rewritten so many times that we had long since reached the point of diminishing returns. He had already approved a much earlier manuscript version and then, a few months later in New Delhi, he had dictated a lengthy revised version to his granddaughter Alexandra. I had gone over all the variations and subtle refinements of the Alexandra version and had incorporated them in the revised text. There was hardly any way it could be improved consistent with his intent. While we were working on a passage quite unrelated, Fuller's eye had casually seen the caption to "The Deliberately Nonstraight Line" while checking the chapter outline. He dropped everything and started working on it from scratch as if the idea had just come to him, as if he had never had a chance to tell anyone about it before, and as if we had nothing else to do. Two hours were being wasted on that last day of our working session as he rewrote the whole story in longhand. I glowered. I crossed my arms. I accused him of self-indulgence. We got angry. We had so much of the book still unwritten that we couldn't possibly waste time rewriting what was already so exhaustively presented. It was the expository reflex syndrome, his sheer pleasure in telling the story again.

Our voices rose. The rest of the family avoided the sunporch. And then finally and unexpectedly he gave in with the utmost charm. For the first time he had been persuaded to drop the nonstraight line in mid-exposition, and he never brought it up again.* I was able to satisfy him that his first version was adequate and that his time could safely be put on something else. But two hours had been wasted and there were still faint whiffs of cordite in the air.

That was Valentine's Day, 1972, and later in the day he asked me for a clean sheet of paper and he drew me a perfectly beautiful white apple with an arrow through it and an accompanying bit of verse. But the poem started off

*Not quite; he tried to rewrite it all again on the galleys.

At the heart
Of the heart
Of the Applewhite
Heart
Are ever
The black seeds. . . .

So he forgave my unmannerly effort to spoil his fun.

Since that day I have gotten so I can see these reflex syndromes coming. I can spot them a few paragraphs in advance. Sometimes I can head them off before he gets fully wound up, but if not, he simply has to retell the story before he can get down to any other work. There is nothing senile about this; it's just the urge to communicate and his uncertainty that he has really told it all before.

In our last working session before the book went to press, we had spent a wonderful October day in the sunporch office. Bucky recalled that he and Christopher Morley could always intuitively sense the approach of a new moon. (His Moon.) He talked of his mother and how she had cautioned him to first see the new moon over the right shoulder. This omen of good luck was a direct link with his boyhood. We worked on until early evening when we had to turn on the light and when Bucky looked up and out of the window, there was the new harvest moon shining over his right shoulder. He said that's mother. Tears streamed down his cheeks. She is sending us a message, the book is good, that's what she's saying. "Just think, Sonny, how difficult it must be for her to get a message through to us, how very clever of her to have done it this way."

7. Cosmic Fishing on the Grand Banks and in the Delaware River

WORKING IN our office porch in Washington was so successful that occasionally Bucky would be moved to celebrate our congenial circumstances. At his work table on 16 October 1972, after pouring himself a fresh cup of tea, he delivered a rationale for what we were doing there—in the form of a nonstop soliloquy. Reading it now, it almost sounds like being geared up for high-speed transmission, like a fast tape for blip broadcasting . . . squeaks and all. The absence of predicates and punctuation results from his delivery, not from my failure to keep up.

Bucky: "Cosmic fish in the Grand Banks that's what's going on in this room where we have developed such a sensitivity that it seems it might be well to stop a moment and take an inventory of what really is going on here there is the business executive type who would be saying why don't we finish the chapter why don't we get this book out but thank god you are not doing that and that is just the opposite of what you're saying letting me do all this . . . what might only seem to be digressions but it isn't important what we read in the newspaper because all the very extraordinarily rapid evolution is going on in the invisible spectrum and the press and TV are really missing the big show . . . what I'm saying is that right now all of humanity is really breaking through to a completely different way of looking at Universe . . . here we have Brendan cutting things out and telling me what is the really

latest going on in physics really beginning to come
out where I've been all along and Brendan and now
all my friends in a sort of new strategy of just
spontaneous deputies associates . . . Ed Schlossberg
very good at this and Allegra really extraordinary at
bringing me things I would have missed so we're all
around getting these reports indicating that science
in general is converging with us and the great
coordinate system of nature as we have discovered
and so what I'm doing when I may not seem to be
working on the book is this really very important
cosmic fishing . . . first you have the intuition and
then you have the second intuition about what you
should do with the first intuition . . . we're coming
in now on really the Grand Banks teeming with all
these cosmic fish that have never been caught before
with comprehensivity of the role of man in Universe
because humans really do have a purpose and the
metaphysical is what is really very suddenly coming
into prominence . . . and these kids really just take
sex and how different it is now when evolution used
to have to reproduce itself and they had to think of
their bodies as just the great baby-making-machine
home . . . well all that's becoming extinct and the
kids don't act that way anymore and the metaphysical is emerging terribly fast and the physicists
all know that I am on to the right thing except that
when that man Teller who went with Robert Oppen-
heimer and the atomic bomb and the hydrogen
bomb and he was talking but the other scientists
present really found me far more cogent and
interesting while Teller is the one physicist who was
just giving the capitalists the big boom they wanted
. . . but now quite clearly we are all coming into
phase so if we make an inventory of what's going
on in this room right now . . . all the permeabilities

of the MITE's and the number of nonregular octa-
hedra and the number of nonregular rhombic
dodecahedrons and then we have next the total
sphere as the convergence in the vector equilibrium
with its spaces and concaves and we have the
concept of the limits of asymmetry in respect to the
vector equilibrium as the limit of coming to the
molecules that's what we have . . . nuclear unique-
ness and all of its variables within the domain of the
three-frequency vector equilibrium and all of the
things we've been doing the past couple of days
dealing with the transformation of the jitterbug and
tensegrity forming from tetra to icosa by sliding the
point of concentrated pressure going from the ends
to the middle and our confirmation of the original
concept that the vector equilibriums are nuclear
structures embracing all the variables of Universe . . .
associating all the molecular build-ups which has to
do with syntax because I am holistic and I really
don't want to be limited by . . . it's like a bunch of
picture puzzles used to have a picture on the box
of what you were making but let's just suppose we
had no picture on the box and we had ten puzzles in
different transparent plastic bags and we mixed
them all up each of the puzzles with the other
puzzles I think we could you and I sort of intuit
which kinds of pieces must be with one puzzle and
which kinds of pieces with another and eventually
we could at least get them all in the right bags again
so they could be worked out . . . and that's what its
been like here working these days like we had a
picture of George Washington crossing the Delaware
and we'd have one little piece that looked like
George's hat and we're really just throwing in the
tiles . . . and that's what I'm doing giving a lecture
when the kids are all following it and you really can

go very fast while you're talking about George
Washington's hat and then you're talking about the
ice around the boat and all you really have to say
and all you have time for in the lecture is just to
say HAT or just to say ICE like that and everybody
follows and we're really throwing in the tiles and
we have the Picture of George Washington Crossing
the Delaware."

8. Spontaneous Deputies

ONE OF THE original schemes for *Synergetics* was for it to be presented in two parts: the first part by Fuller and a second of contributions from his students and contemporary colleagues. The scheme was never abandoned, but it was only partially realized with the incorporation in *Synergetics* of the major contribution of Dr. Arthur L. Loeb. I was not qualified to edit any of the contributions and we were unable to find anyone competent and willing to prepare them for publication.

> *RBF: They didn't need any preparation. I found them more fully prepared for the publisher than any of the manuscripts I have turned in.*
>
> *EJA: Probably so. But each of the contributions was prepared independently, with no reference to each other, with widely varying subjects, assumptions, technical criteria, academic apparatus, and illustrative approaches. None of them had specific references to the basic text—if only because no such text was available at the time they were prepared. Without some coordinating theme there would be little rationale for binding them between the same covers. Michael Burt's contribution was over two hundred pages. Anne Griswold Tyng's has since emerged as a doctoral thesis.*

Fuller thinks I put too much stress on the need for an editor, but I felt the contributions could hardly just be wrapped up in manila paper and twine and sent off to a publisher. Furthermore, the book to which they relate had been completely transformed since the contributions were originally written. Some of the original student projects have been superseded by independent publications, notably Keith Critchlow's

Order in Space (New York: Viking Press, 1969) and
Anthony Pugh's *An Introduction to Tensegrity*
(Berkeley: University of California Press, 1976),
both of them dedicated to Buckminster Fuller.

All but one of these synoptic papers were from
Fuller's design students who had been inspired by
his innovations and who had themselves contributed
at least to the articulation of his geometry. In some
cases—Kenneth Snelson, Donald Richter, Ted Pope
—they had been responsible for further original
discoveries. This is a phenomenon that Fuller has
described as "coincidental articulation."

RBF: Conceptual momentum.

*EJA: The interaction of the individual and the
group . . . neighboring disclosures, spontaneous
deputies, and coincidental articulations.*

He felt a strong sense of obligation to see that their
contributions got appropriate recognition at the same
time his own major work would be published; he felt
their concurrent publication would have an enhanced
synergetic effectiveness. At the same time, he was
concerned that his unique discoveries should not be
released prematurely in the partial contexts of
specialized papers with limited focus—and perhaps
inadequate recognition of his central role. His
assumption of responsibility for the publication of
the papers had its preemptive aspects. It is significant
that he refers to them as "neighboring disclosures."

Arthur Loeb was the only one of Fuller's potential
collaborators in the 1960s who enjoyed a widely
recognized academic reputation. He had been an
associate professor of electrical engineering at M.I.T.
and was later on the research staff of the Kennecott
Copper Company's Ledgemont Laboratories in
Lexington, Massachusetts; eventually, he joined the

faculty of the department of visual and environ-
mental studies at Harvard. Loeb's principal work as
a scientist is in the field of crystallography. He is
also an amateur musician and past president of the
Viola da Gamba Society of America. Fuller and Loeb
first met under the fortuitous auspices of the Boston
public television station WGBH-TV, where Bucky
had often performed since the pioneer videotape
programs of the early 50s and where Arthur was
known as a musician.

Obviously, Loeb was never in the same category
as the students described earlier. His completely
independent stature derived from original research
and publication.

*RBF: Loeb's original work had commenced, however,
circa 1954—two decades after Fuller's 1934 discovery
of the convergent-divergent, rational-volume hierarchy
of geometrical geosystems, commencing with the
tetrahedron, which Fuller had first published in 1944.*

Fuller's preoccupation with priorities causes problems for his many friends in academic life. Arthur
Loeb has written me, "What was momentous about
our meeting was that for entirely different reasons
and working with structures differing in scale by a
factor 10^{10}, we should have hit on the same natural
forms. . . . The forms have been known for eons; it
was what both of us have done with them that is
significant. In the same vein, I was indeed surprised
that Buckminster also knew the 3:1 ratio of the
volumes of a parallelepiped and the inscribed
tetrahedron. Whereas I had calculated the ratio
for myself, I was under no illusion that this was
in fact an original discovery. I just knew that these
fundamental relations are in these times not part of
our 'lore.' "

That Loeb's contribution survives in the published

version of *Synergetics* is a tribute to his commitment to Fuller, his forbearance with interminable delays, and his willingness to venture so far beyond the academic precincts.

While spending the New Year's holidays with his daughter Allegra in Pacific Palisades, California, Bucky dictated to Anne a letter to Loeb inscribed by her with firm cursive characters in black ink. (Both Anne Fuller and Arthur Loeb are accomplished calligraphers.) The letter, dated 6 January 1967, says in part:

Dear Arthur—

My thanks for your letter and your report on the vector equilibrium. I am simply delighted with it.

I am also delighted that you will do the preface for my book. That is well underway at last. I have gotten a Graham Foundation grant. . . .

As well as your preface, my thought is to have colored pages, or a separate typeface inserted throughout the text where you will want to make a comment. You will have the completed manuscript at the earliest opportunity so that you may digest it before your writing.

There is excitement in the air as we undertake this last phase of a fifty-year search for re-bridging the gap between Science and the Humanities. . . . The world of science . . . is not going to take quickly to our new tools—so you are the bridge-builder from me to the rest of science. I would like you to say what I say . . . to the scientists in their language. Our terms often coincide but are ambiguous, and, at times, contradictory. I think you can smooth out the differences.

I am full of joy to be working with you!

My very best to you,

Buckminster Fuller

The plans were refined to have Fuller's words printed in red, and Loeb's in green, or vice versa. Perhaps anticipating the vicissitudes that *Synergetics* still had ahead of it, Loeb managed to organize his contribution for easy reference without the close integration that would have been preferable and would have accorded with the original design. It was impracticable to provide Loeb with a firm copy of Fuller's text until it was in galleys. These problems would only have been compounded if we had tried to give comparable treatment to the neighboring disclosures from the students.

One of Fuller's most endearing traits is a genuinely gregarious disposition and his spontaneous outreach for companionship. If he is going out to lunch, then the more the merrier.

> *RBF: Not so. . . . But I am continually being asked by my friends if I would mind if they brought so-and-so along, and because of my wish to sustain the friendship I say "yes" and being brought up as I was, I say it as though I were enthusiastic about the idea, rather than condescending.*

And so it is in writing: he likes the idea of people working together as a group. Thus Fuller regarded it as highly auspicious that we had scheduled a joint working session at his beloved Bear Island for the third week of August 1971. There is no question that he finds Bear Island the most congenial and stimulating place on earth; the sentiment of the family traditions, the nearness of the stars, the day-long racing in the sloop, combine to afford Bucky an annual psychic rendezvous of intimacy with the universe. I had brought up a hundred pages of manuscript and my portable typewriter with some misgivings—because I knew there is never any privacy on an island—but with curiosity in seeing how we could function in what he regarded as such ideal circumstances for writing together.

The first night everyone, even Bucky, was exhausted after a day of sailing.

RBF: After half a century of having the slowest boats in Penobscot Bay, Intuition's winning the crown of Penobscot Bay sailors—including the nation's "best"—was a very special event. Anything but daily routine.

Your description of my sailing gives the impression that I am just a philanderer and pleasure seeker. That was a very unusual week of the New York Yacht Club races culminating in the annual race around Deer Isle. The N.Y.Y.C. was ending its summer schedule in East Penobscot Bay, where I had raced all my life but only with makeshift, not really competitive boats. This was the first time in my life that I had a first-rate competitive racing machine in Intuition. It was the only week I ever did any racing at Bear since I got Intuition so you give all the wrong impression.

EJA: Of course, Bucky, you are right. Not being a sportsman, I even meanly forgot to say who won the race. Everyone on Bear that night was thrilled and talking of nothing else. It's just that I am trying to write a book about writing, not racing.

This recalls your distress when you saw the first proofs of the dust jackets for your book, Intuition. There was a color photograph of the bow of the boat "crossing the finish line at New Bedford." Nothing about who won. But you had added so much new verse to the book after it was set in galleys that they had to add 30 pages and thus raise the price. For this reason they had to discard the first run of jackets and I was able to get the caption changed to read "first across the finish line at New Bedford."

I suggested that we should wait until the next day before trying to get any work done. But he wouldn't hear of it. The kerosene lamp was turned up in the Eating House and as I spread the manuscript pages out on the dining-room table the crowd began to gather round. The Fullers' handsome grandchildren, Alexandra and Jaime Snyder, crewed the sloop and spent the whole month on the island; with them were an equally attractive group of about a dozen high school and college classmates comprising Los Angeles' easternmost suburb. They were fascinated to See Bucky Write. He held manuscript and pen in hand, but it was the irrepressible flow of talk that had the young people entranced. His discourse had only the most oblique relationship to the problems of the chapter on which we were working, and when the young people brought in pancakes at midnight, I was no doubt unduly concerned at syrup getting on the canary yellow pages.

Bucky was presiding among the people he loved best in the close and heated atmosphere of an English coffee house. To him it was the supreme act of writing, the verb, the event. I had heard Hindus describe how the act of sweepers ineffectually combing the temple steps with thin straw brooms was more significant and moving than whether the temple was actually clean. With this thought in mind, I tried to see the act of writing in a new light.

> RBF: Quite the antithesis of my concept. You know I work and rework until I have become certain that I am both comprehensively and incisively at the condition which defies further inclusive and synergetic significance-harvesting.
>
> When the kids came to the Eating House to watch me write they were always very respectful and just sat in silence. You say I am gregarious. . . . Well, okay . . . but I am used to being very alone in a crowd. I like to know that I am related to humanity while I really concentrate, so that's why they gather with me in the Eating House and they just sit really quietly.
>
> EJA: Well, no two witnesses see an accident the same way. I certainly do not suggest that Bucky is playing

*idle games or being a dilettante. For Fuller, having
the young people around him and talking about
synergetics is like giving a lecture, an essential
stimulus and an end in itself. But for me, with my
single-minded concentration on the manuscript, the
words of the text, any session like a group discussion
or a lecture is a waste of time if it does not result
in new material for the book. For Bucky, these
group sessions were very hard work, worthwhile in
themselves because they brought to his creative self-
discipline the energy of a public act.*

For Bucky it must have been a great disappointment that I was so oblivious to the charms of the
place. In fairness I must confess that I am a
confirmed claustrophile: urbanite at heart, sedentary
by nature, a lover of the great indoors. I could not
understand how an island of 15 acres

RBF: 87 acres

could afford no privacy where we could work alone.
Bucky could not understand why I wouldn't want
all the rest of the people to join in and help us with
our book.

RBF: I did not assume any help to be forthcoming.

It took me two days to slip away with my typewriter
and papers back to the mainland,

*RBF: Deer Isle (five square miles) with causeway to
smaller island and bridge from latter to mainland.*

as I had neglected one of the cardinal precepts of
an earlier training: Never get on another man's
island without your own boat.

There was a family house on Deer Isle which,
aside from telephone and running water—unavailable
on Bear—afforded some quiet and comparative
freedom from distraction. I decided to wait there in
hopes that a stormy day would keep Bucky off his
sloop and give us a chance to stay indoors and

115

work on the book. One of the regular old Bear Islanders whom I had never met before came to Deer Isle the next day with a message. He said, "I don't know what your relations with Fuller are but unless you are prepared to be totally independent of him in the future, then you sure better get the next boat back to that island." I did go out to see Bucky to explain that I didn't mind hanging around on Deer Isle, but that it seemed hopeless to get anything done in the consuming pattern of family life on Bear. It was all amicable. All we had to show for the week was two brief work sessions at the Deer Isle house, rewriting about four manuscript pages. We could have done as much by telephone.

The fiasco of Bear Island did not mean that Fuller had abandoned the lure of group composition.

RBF: No! I don't have any thoughts of group composition. I never wanted to even try group writing. I just wanted Arthur Loeb and the others to help articulate my half-century of exploring. I've always been so busy that I've never had enough time to do everything myself, so I wanted them to take over . . . but I wouldn't expect them to do any of the writing.

EJA: Fuller is right in objecting to my phrase "group composition" because he did not necessarily expect his collaborators to put words on paper. Getting things on paper is not one of his major preoccupations. He does like a collaborative team for mutual stimulation and for the talents they could bring to bear—whether as people well versed in physics or chemistry, as draftsmen or mathematicians, or merely the attentive and curious young. Fuller wants such people to help him refine the statement and exposition of his philosophic and design concepts— but without the expectation that they would sit down and write a report like a committee. Anything that had to be written, Fuller himself could write: all he lacked was the time.

> *Most of our working sessions would begin with our*
> *just talking about synergetics; since I wasn't a*
> *group I could do a first draft and Bucky would*
> *rewrite it.*

Far from it. The woods were full of potential collaborators. There was a bright and beautiful girl in Montreal named Michele Cuevas, and he wanted me to get in touch with her as he was sure that she would be of great help to both of us in writing the book. Tony Huston was in London working on passages from "Numerology" and the Naga theme; his talents should be brought into play. Brendan O'Regan had studied physics in Dublin and tentatively joined our writing project, but he was highly itinerant and contributed mostly by correspondence. Only the cruel circumstances of chronology prevented his grandson Jaime from playing a more prominent role in the book that he might have done had he been through college.

> *RBF: This is misinterpretation of my thoughts and*
> *words. I did not seek out these people. They sought me*
> *and I thought of them only as being Spontaneous*
> *Research Associates.*

> *EJA: I have failed to interpret RBF in a way that he*
> *can recognize. Perhaps part of the problem is that for*
> *RBF—ideally if days were a week long and if he could*
> *start all over again—it would all be one single book:*
> *the dome engineering, the Naga theme, world game,*
> *cosmic accounting, geometrical hierarchies, they are*
> *all integral, one of a piece; only the rigid, arbitrary,*
> *and accidental restraints of publishing economics seem*
> *to have forced publication of the books as separate*
> *packages.*

A year after our Bear Island workfest, he received in his Carbondale office three curious and unsolicited letters from graduate students in widely separated institutions. They dealt with unorthodox modes of

geometry and suggested certain correlations with
Fuller's synergetics. He dictated a letter to me about
this in the St. Louis airport, dated 17 October 1972:

Then we had to move down to the gate, and Bucky
continued dictating, this time speaking directly to you. . . .

The arrival of all three of these papers in my hands
simultaneously points up very powerfully the need for
having Arthur Loeb almost continuously in close
association with us for reading and correlating our
work. I do and will continue to think extremely well of
Brendan, but think Arthur is in a maturer position and if
the two could work together this would be highly
synergetic.

The moment we have funds, the institute should go
into bringing these two together. In due course, I would
like to bring Ed Schlossberg into the intimate relationship
with them. I think they are all capable of spontaneous
cooperation with each other as well as myself. I am
eager for you to be thinking about this and for you to be
ready to take constructive measures in those directions
when the opportunity arises.

It is truly thrilling to see the book taking shape at last.

The receipt of such broad marching orders on the
third anniversary of our collaboration might suggest
a certain fragility in the tandem character of our
working relationship. The obvious goodwill and
enthusiasm of Bucky's proposal for a renewal of
group articulation (since I may not say group
composition) led me to suppress my practical
objections to it. I did not share Fuller's sublime
indifference to the constraints of earning a living,
and I could not invite others, however willing and
talented, into arrangements for which I saw no
prospect of financial support. Meanwhile, volunteers
would always be welcome.

RBF: Sonny, by this time—1972—I had written and
published ten books and thousands of articles. In none
of them did I employ any group writing technique

and I am confident that I have never conceived such a phenomenon to be possible.

What I have cared about was the cultivation of artists who showed any comprehension of synergetics. I have had a half-century of little or no comprehension of en-syn-geom . . . and I had a strong intuitive urge to hold these people's interest.

It is necessary to think about the half-century of the synergetic-math search for nature's most omni-economical coordinate system. . . . The original discoveries of finding them years ago. . . . Then the counsel by Thornton Wilder to hold off publication. . . . And then the sudden rushing in of the students with their impatience to publish. . . . I was dealing intuitively with nature's coordinates and I did not know yet what shape the reduction-to-practice might take. I was trying to forestall further publishing by others, such as Critchlow who, despite the dedication to me, did not say that I had taught him the whole idea of the rational hierarchy, etc.

EJA: Bucky's affinity for artists and his encouragement of allies are both testimony of and compensation for his loneliness—a loneliness resulting from the failure of others to understand synergetic geometry.

Fuller's objections are reinforced by three of his ardent convictions:

(a) that the decision to postpone the publication of his synergetic mathematics entailed great risks but was fully justified;

(b) that the impulse of students to rush into print with only aspects of the geometrical hierarchy was subversive of his highly self-disciplined strategy; and

(c) that his timetable of philosophical exposition would have been better served if all the original spontaneous deputies had continued to stay on the team to function as a collaborative group dedicated to the original RBF strategy.

If Fuller had the means to underwrite his ideal of group cooperation, I could envision a permanent floating newspaper city room—like The Front Page.

We could all function as what he calls "spontaneous deputies," writers with green eyeshades and sleeve garters, physicists with pocket calculators, and artists over drafting tables, all untrammeled by sordid thoughts of personal by-lines or Friday paychecks. Fuller's ego does indeed have this plural aspect congenial to the prospect of lieutenants overcoming all the frustrations of the things he never has time to do himself. His role would become partly that of the mythical figure of the Great Pirate—dividing and conquering, dealing with each individual talent on a one-to-one basis—and partly that of Danny Kaye as Walter Mitty with hints of megalomania.

RBF: Just not so.

EJA: I have sketched what Fuller calls a "cartoon." The trouble with spontaneous deputies is that it is hard to sustain the spontaneity. There is a point at which the Critchlows and Snelsons wander off the reservation. This is what I am trying to suggest.

9. How They Brought the Good News From Carbondale to Third Avenue

FOURTEEN YEARS BEFORE *Synergetics* was published, Fuller had sold the rights to the Macmillan Publishing Company for an indecorous advance of $2,500. The book was "to contain approximately 70,000 words . . . based on recent lectures." It was a perfectly good contract reflecting what was then fairly considered the best interests of both parties. But by the time the agreement was 10 years old, Fuller's circumstances had changed markedly. His other books had earned an enthusiastic following and he was becoming more widely recognized as something of a world figure.

Fuller's office actually maintains a quantified record of his increasing recognition and public activities, compiling annual totals of the accelerating upward curve of articles by and about him, degrees awarded, newspaper articles, audiences lectured, television talk shows, radio interviews, thousands of miles traveled, and so forth. There seems to be no limit to the extent such matters can be measured: three pages in *Britannica III*, the longest entry in *Who's Who*.

> RBF: *This is not an ego matter. . . . Just my record as a guinea pig. Just as a navigator has to "take sights," and "work his position," and "keep log books." There is no ego involved. It's just a matter of attempting to set in order the facts of experience.*

His lecture schedules and architectural commissions left him less time than ever for writing. There appeared to be no way he could afford the time to finish a book of the dense immensity of Synergetics.

Fuller eschews agents, in both strict practice and broad principle. He turns down all overtures from lecture bureaus. "I do not promote," he says. "I have a basic discipline: Never ask anyone to listen. Speak only when you are asked to speak. When that occurs give them your best."* You don't need an agent for that kind of strategy.

Like his lectures, his book productions had a correspondingly impromptu character: eight books involving eleven different publishers—an aggregate of ad-hocnesses. So much for agents, but when it comes to doctors and lawyers he has respect for profession. Since the 1930s, he has had highly talented patent lawyers; one of them, Donald W. Robertson, wrote a book about being Fuller's patent attorney.† An old-line Federal Street law firm in Boston handles his dome royalties. By 1969, when I had made my first visit to Carbondale, it seemed the time had come to seek out competent legal counsel on literary property matters. Bucky authorized me to explore this unfamiliar field and make recommendations to him. And so we found in Gerald Dickler of New York City an accomplished authority in the field with the temperament and tolerance to accommodate the unpredictable interests of an idiosyncratic client.

Dickler consolidated Fuller's various titles in print with a variety of university presses and technical houses in one wholesale contract with Doubleday-Anchor. In renegotiating the Synergetics contract, Dickler was able to identify and serve the interests

*RBF letter to Bruce Carrick, Macmillan, 17 September 1974.
†Donald W. Robertson, Mind's Eye of Buckminster Fuller. New York: Vantage Press, 1974.

of publisher and author better than either could have done by themselves. On Macmillan's side of the picture there was a perfectly valid contract, but no reasonable prospect of fulfilling it unless time could be bought so that Fuller could do the writing at the expense of time otherwise devoted to revenue-producing assignments like lecturing and architectural commissions. It was frustrating for Macmillan to have waited so many years on an author unwilling or unable to release the manuscript. A succession of new editors at Macmillan had left the author with a waning sense of commitment; we wanted to refund the advance and buy up the contract to gain the freedom to start afresh with a new publisher. Bill Whitehead was then our editor at Doubleday-Anchor; he had a sympathetic grasp of Fuller's philosophy, and we wanted to show the manuscript to him. These were our instructions to Dickler, but he advised against such a course, both on legal grounds and for marketing prospects: he felt that Macmillan was the right house, one of the few that could give *Synergetics* the in-house production care, promotion, and distribution, to successfully launch such an unconventional work as a trade book.

By early 1972, Fuller and I had completed the first third of the new manuscript for the projected work. Prior to submitting this increment to Macmillan, Dickler proposed a new contract to replace that of 1961. The operative effect of his proposal was that it would cost Macmillan $25,000—as a minimum commitment—to see this first third of the manuscript. (This was a little like making them pay to see their own hole card, but they were realistic in recognizing that no book would be forthcoming on the basis of the original advance.) Dickler's new contract proposed a schedule of a $25,000 advance installment

123

with each successive third, for a total advance of $75,000. But there was a rather unusual special provision in the proposed contract. If the publisher determined, after seeing the first third of the manuscript, "that the work does not indicate that it will recover an advance of $75,000 . . . he may reduce the offer." But any reduction by the publisher below the full amount would release the author to "submit the project elsewhere."

For Macmillan, it was like a blind auction. For the author, the ingenuity of the proposal was the assurance that Macmillan would remain committed to publishing the work—at a hypothetical lower advance—if the author was unable to better the offer with another publisher. As it turned out, Macmillan committed the first $25,000 to see the first third of the manuscript. At the end of the 60-day period for its inspection, they decided to commit the full $75,000 to the book and thus foreclose any possibility of losing it to another publisher. Dickler had foreseen, correctly, that this substantial advance would be sufficient compensation for our disappointment in not taking the book to Bill Whitehead.

This greatly increased advance was also an essential element in reactivating the author, while Macmillan was protected from the very real prospect of losing to another publisher a work for which they had waited so long. The new contract proved to be the crucial instrument in bringing the work to completion. The payment of the advance in three installments as each third of the manuscript was completed nicely achieved its calculated effect. In fact, the final third was submitted nine months ahead of the contract deadline. (When we had wrapped up that last installment I brought it up to Manhattan in an old shirt box, delivered it to

Macmillan, and headed for Gowanus Basin in
Brooklyn where I boarded the Polish freighter
Zambrow for an eleven-day passage to Hamburg.
I had plenty of time to read the *Idylls of the King,* as
only the captain spoke English.)

But there were to be many pitfalls ahead, and
the need for Dickler's guiding hand was by no
means ended with the signing of the contract or the
submission of the manuscript.

There had been resistance at Macmillan to the
idea of associating Dr. Loeb with the book. They
had no misgivings about Loeb, but they wanted no
name other than Fuller's. Loeb had considerable
proprietary material tied up in his contribution,
material that he could easily have incorporated in
books of his own that were in progress. The real
possibility that further delays might give Loeb every
excuse to withdraw was a critical factor prompting
Fuller to concentrate on meeting the contractual
deadlines. Here again, Dickler was instrumental in
working out a solution equitable to both Loeb and
Macmillan and consonant with Fuller's promises to
Loeb.

When the final manuscript was submitted to
Macmillan in March 1974, *Synergetics* quickly found
its first audience at the shirt-sleeve copy-editor level.
We were fortunate in being assigned Ken Shaw as
copy editor and Maurice Schneps as typographical
and layout designer. Shaw's first love was music
but he was broadly knowledgeable in other fields
and—though skeptical of its pretensions—
sympathetic to the book's strategy of embracing
such a variety of disciplines. He was diligent in
enforcing consistency in my rather complex system
of numbering each section of the text and resourceful
in integrating the text with the illustrations. Copy

editors are incurable perfectionists: we had submitted a plate from the voyage of H.M.S. *Challenger* (at Fig. 203.09), but the quality was not good enough for Ken, so he went down to the New York Public Library and brought back a cleaner impression. He was a stickler for grammatical conventions; time and again he would discover paragraph-long sentences that I had launched without a functioning predicate—maybe a handful of gerunds, but no predicate. Another stumbling block was Fuller's peculiar capitalization, and on this score I had to fight hard to prevent Shaw from revoking Bucky's poetic license. The high quality of the final text also owes much to Macmillan's fiercely independent chief of copy editing, Miriam Hurewitz.*

Maurice Schneps designed the book with a discerning eye to Fuller's philosophic and artistic intentions. He frequently was called upon to improvise compensations for all the widely varying styles, sizes, qualities, and sources of the illustrative materials. When the book finally went to press at the Vail Ballou Press in Binghamton, N.Y., Maurice virtually went to bed with it; he personally checked the press sheets for registration, printing quality, last-minute corrections, and other production refinements. Both Maurice and Ken were long-suffering with our enormous changes at the last minute; their talents greatly enhanced the final

---

*Hurewitz and Shaw engaged John Berseth of Princeton, N.J., as outside copy editor; though we have never met, he earned enormous respect for the quality of his blue-slip queries and his impressive span of attention spread over a thousand manuscript pages. Cynthia Insolio proofread the first half and Henry Engel the more intricate second half of the first set of galleys. The second set of galleys was read aloud to each other by Hyman and Rosetta Jacobs—I would like to have heard that.

result, and throughout that long winter they were the only ones to give us a feeling that someone at Macmillan knew what the book was about.

Fuller was not involved in the first stage of production, the copy editing of the manuscript. He was also insulated from the later stages: the page proofs (where the corrected galleys are divided up into pages), the "repros" (the proofs that are photographed for the making of the offset plates), and the "blues" (which are used to check the quality of the camera work). It was in the intervening stage of the galleys that his creative urges were given free rein.

In October 1973, the first galleys began to trickle in, and the effect on Fuller was galvanic. There is a convention in the publishing world that a work is substantially complete by the time it is set in galleys. Everything but the final page numbering is supposed to have been determined. The author has the privilege of a brief period of examination to check the book as set in type for conformity to his manuscript. Any errors in the galley not appearing in the manuscript are corrected at the expense of the publisher. Any of the author's alterations of fact, style, spelling, or grammar— above a standard allowance—are chargeable to the author. Changes of substance, new material, additional content, are generally not contemplated after the galley stage of production.

It is, of course, always highly gratifying to a writer to see his words finally in print. The printed words take on a wholly new character—a life of their own—and there is a compulsion even for their own author to read them as if he had never seen them before. This is what Fuller did, but in his case there was nothing final about it.

Cosmic Fishing

RBF: When I met with Bruce Carrick, our editor at Macmillan, in October 1972, I explained to him in some detail about my method of composition . . . how a thing is clear when I first write it down on paper and how much clearer it is again when it comes back after you [EJA] have typed it up. But you really can see vastly more only after it has come back to you as galley proofs. That's when the real writing can begin. This is not a question of being careless. I explained about meeting Winston Churchill's secretary at the New York World's Fair and how we both had to go through seven drafts. That is the ideal, seven. Churchill had his speeches and other drafts set up double-spaced in cold type. With me I use galleys. My use of galleys is not facetious: it is part of the process of being adequate and thorough.

I had also told Fuller that James Joyce had achieved some of his finest effects in unbridled corrections to galleys. And why was he not to profit from such exalted examples?

RBF: I did not know about James Joyce. I was not emulating anyone. Glad to have confirmation of similar needs and tactics of others.

There was nothing capricious about Fuller's feeling for galleys.* Editors, accountants, and lawyers would join me in stern and unfeeling warnings about the horrendous prospect of overcharges for author's alterations. The only thing that mattered for Fuller was integrity of expression, and if it was to cost him over $3,500—which it did; the book had to be completely reset—that was just an unavoidable charge to the cost of an unfettered muse. We weren't a third of the way through the galley corrections when Carrick and I decided that we could in no way let Bucky even see

*Except when he would use the galleys for instructions about airmail postage or reminders to turn up the thermostat or turn off the hot plate.

the page proofs or the whole process might be started up all over again. We made no secret of this. It was charming and characteristic of Fuller, in an access of self-knowledge, to be in complete accord.

RBF: *You forgot to say that I explained all this to Macmillan ahead of time and that they broke their word about accommodating my methods.*

EJA: *Strictly speaking, that is true. They had promised some kind of print from the computer tape from which Fuller could rewrite in advance of the galleys. This was an experimental mode, but one that would permit corrections without alteration charges to the author—if we could manage to cope with the "idiot tapes" in computer characters rather than type. Fuller and I understood that this print-out would not look like or feel like a galley and that we would have to acquire a new computer baby talk to cope with it. This we agreed to do.*

What went wrong was that computer typesetting at that time was brand new, far from perfect, and unfamiliar to Macmillan. The situation was aggravated by faulty coordination between the editing and production departments. The particular composing system chosen would not produce the necessary intermediate steps between manuscript and galleys.

Galleys galvanize Fuller partly because of the large visual component of his imagination. The effect is reflexive: his imagination is triggered by what the eye frames in front of him. It was the same with manuscript pages: he never liked to turn them over or continue to another sheet. Page = unit of thought. So his mind was retriggered with every galley and its quite arbitrary increment of thought from the composing process.

When Fuller is through with a galley, it often looks like an untended switchboard with all the wires plugged into a margin of floating addenda (see page 130).

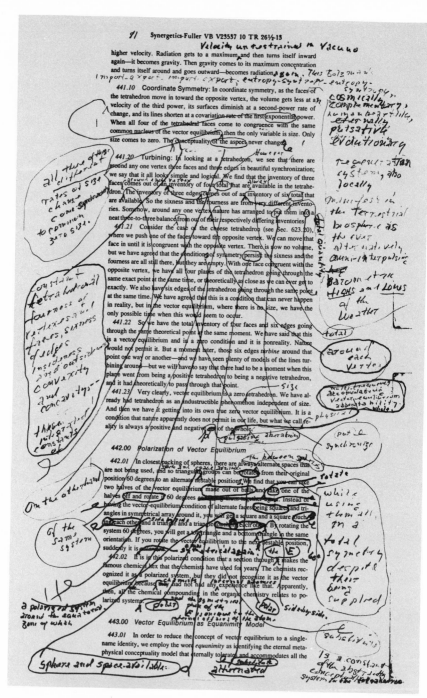

Fuller's galley corrections to secs. 441 and
442 of *Synergetics*.

We worked on the galleys in intimate communication but usually a continent apart. Every other mail would bring a new batch of galleys postmarked Prince Edward Island or Pullman, Washington. I would transcribe them into what was in effect new manuscript. As in the original composition, there were few corrections or deletions. It was all new content; sometimes nuances of style but more often matters of substance. Once, working together a whole day at the Beverly Hotel in New York, we completed only two galleys between breakfast and 11:00 P.M.—one of these is shown on page 132. With a fine pen using red ink and green ink he had drawn an exquisite sketch of a spherical triangle subsiding into a planar: he was anxious to get that into the book. Anyway, at midnight we went across the street to the Waldorf to celebrate with a steak dinner. There was a pretty nightclub photographer with a flashbulb camera. Always one to convert what others might consider only an occasion into a matter of destiny, Bucky insisted that we have our pictures taken. Fuller is proud of his skill as a photographer, so first he had to instruct the young woman—politely but firmly— in her chosen profession. The eventual picture came back from the darkroom as a fair likeness featuring his credit card and my martini cocktail in the foreground. He inscribed it to me affectionately: "Entering the home stretch of the $\frac{1}{2}$-century-long synergetics galley race. B.F. and E.J.A. Jr. 'At the Waldorf,' January 9, 1974."

For Fuller the greatest function of the galleys was to concentrate the mind, as Samuel Johnson said of the death sentence. Throughout his career, Fuller had been encapsulating his mathematics in cryptic notes and private equations—never

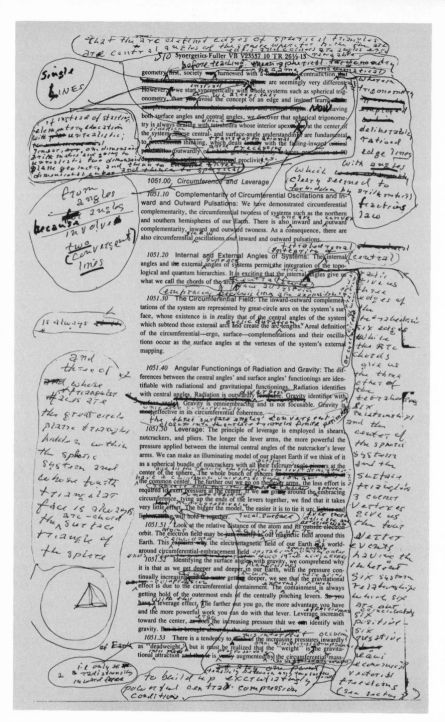

Fuller's galley corrections to Sec. 1051 of Synergetics.

conforming to common usage—caching them away
to document his insights, but not really
communicating the total geometrical hierarchy.
Even writing the book, I felt there were times when
we were recarving the Rosetta stone of his
inscrutable 1944 memorandum. It was the galleys
that provoked him to reveal connections and
relationships and insights that he had long kept
secret. He would work with the desperate intensity
of a canoeist too close to the brink of the spillway.
I watched the progress from the distant shore—
transfixed but powerless to intervene. Here he was
really marking his own paper.

The galley confrontations prompted several
prodigal creative breakthroughs or, as he called
them, intuitively inadvertent cul-de-sacs. All of the
book's climactic definition of his hierarchy of
concentric geometrical hierarchies was articulated
in white heat as a sort of space walk of 24 new
manuscript pages from three pages of galleys,
starting in mid-December 1973 in Washington
and culminating in feverish composition between
Christmas and New Year's Eve in Pacific Palisades,
California.

RBF: *Sonny, this concentric hierarchy was conceived
from the outset and as early as 1934. But I had one
value more to establish, the volume of the sphere of
unit vector radius, which proved out as five—in 1973.
I had made a model of the concentric hierarchy as early
as 1947, but this was the first time I had tried drawing
it.*

EJA: *The conception was earlier; the articulation was
at the last minute.*

These forays from the galleys contain his first explicit
formulas for the area of a circle in triangular
modules and for the volume of a sphere in

tetrahedral modules—all without *pi;* they are to be
found in sections 982.50 through 985.22, ten pages of
new text through which the outlines of the original
galley text can be only dimly reconstructed.

> RBF: *The correction factors used I found and included
> in the chart of the hierarchy which I sent to the
> Raleigh Group in 1947. It was first published formally
> in* No More Secondhand God *in 1963.*

The text describing the geometrical relationships
of tetrahedra and spheres was set up in the three
galleys he worked from, but it was extremely terse
and obscure. I had typed the manuscript from a long
holograph he had written in the hospital in St. Louis
the year before (December 1972). I knew he had
not been totally satisfied with the solutions he had
worked out in the hospital, and if I had asked him
whether they should go into print, he would probably
have said no. Though they were not clear to me,
they were the best statements I had had from him
to date on the subject, and I submitted them as part
of the manuscript. Thus the galleys functioned as
cryptic notes from himself to himself over the
intervening twelve months.

This is an extravagant way to write a book, but it
proved to be effective. At the time, he wrote me an
explanatory note attached by Scotch tape to the
almost indecipherable galley: "Sonny, if I had not
entered this intuitively inadvertent cul-de-sac I
would never have been excited—by attempting to
correct it—into discovery of the neat 5-volume of
the nuclear sphere, which eliminates the necessity of
employing *pi* in synergetical coordinate systems,
though it discloses where and why *pi* coexists, but
only as a terminal vestige."

Other major passages in the book were written in
such gestalt bursts between galleys; examples may

be found at sections 905.00 through 905.73; 930.11 through 930.26; and 1053.13 through 1053.17.

I couldn't help suspecting that all these last-minute outbursts must be part of the pattern of Fuller's lifelong preoccupation with the secrecy of mathematical knowledge: how the navigators and high priests jealously guarded their private means of calculation, reluctant to surrender their source of power and the source of the king's dependence on them.

There is a scene in *Animal Crackers* where the other Marx Brothers berate Harpo for having stolen the table silver at Margaret Dumont's dinner party. They lash him with mock abuse and he responds with equally spurious remorse as he lets small cascades of spoons and butter knives fall in penitence from the stuffed sleeves of his raincoat. In a final gesture of abjection he lets drop the whole silver coffeepot. I still have not digested all of Bucky's whole grand package of add-ons to the galleys, but somehow it reminds me of Harpo's coffeepot.

10. Buried Alive

PEOPLE FAMILIAR WITH the number of years
Bucky and I had been working together—friends,
cousins, creditors, in-laws—would often ask me . . .
What does Macmillan think of your book? They
envision publishers, unlike banks or utilities, in highly
personal terms. They presume publishers to be rich
and cozy, half Dutch Uncle, half Mother Earth. I
found it difficult to explain that, though I had a lot
of good friends at our publishers, to ask how
Macmillan feels about *Synergetics* was rather like
asking A & P how they feel about lamb chops.

I have an imaginary picture of a publishing house
as a loose association of people, many of whom
have just come from, or are planning to depart for,
other publishing houses. The editors are committed
chiefly to their overlapping responsibilities for books
in progress. The business people are on the top
floor, and they make all the decisions about book
after book and writer after writer: whether to
publish, size of advance, price, promotion, when
to let a title go out of print—everything. Most
business decisions are made on projections of
numbers, and since books cannot be easily quantified
(except as accessories after the fact), publishing
is inherently a very speculative business. Thus the
businessmen become dependent on the persuasion,
track record, and artfulness of the people they
employ as editors.

Editors are artistic and sensitive like writers, but
they understand business—which writers don't.
The editors are employees only in the sense that
they can be fired. (Whom did Brecht have in mind

when he asked: Why murder a man when you can employ him?) I have the impression that editors function as free-lancers in a bullpen—like people on the trading floor of Lloyds of London or the New York Stock Exchange—making book, with their own market sense but with their boss' money. Wrapping up a package.

Editors seem to have to cope with as much in-house chaos as out-house chaos. At least this picture explains my experience that those in a publishing house best able to understand a writer's problems are those usually least able to do anything about them. (A situation of that kind could hardly be accidental.) And when you do get into your editor's office, you find there are never any tables and there is no place to put any of your papers down. You have to work out of your lap, which is even worse because the chairs are so low.

Just before publication of our book, the situation was as awkward and precarious for Macmillan as it was for Fuller—though we all seemed oblivious— at that point of no return after fulfillment of the contract and the exchange of the bride price. It was too late for any thoughts of incompatibility or antipathy. For Bucky and me, Macmillan was the ultimate nexus, the only possible link where his design-scientist's thoughts could reach print in the industrialized capitalism of a mass society; there was no other kind of patron available. Only in retrospect is it clear that there was no way the gears could be expected to mesh smoothly. How could a committee on tenure generously receive a Pythag-oras? What use could a Boeing technical task force make of a Copernicus? How does Buckminster Fuller shape up on the fall list? For both parties the

shock ahead was mutual, inevitable, and totally
unexpected.

Once the galleys were locked up and the book was
sent off to the printers, it would have been nice to
think that most of our troubles were over. *Synergetics*
was just the kind of work that is almost impossible
to appraise from the manuscript: the span of
attention demanded by its endless intellectual
labyrinths, the complex relationship of text and
illustration, the exotic scope and terminology, the
sheer bulk, all made it very difficult for the publisher
to assay his property. The book opens with an
exhortation to the reader to *dare to be naive* . . . but
it is quite another matter to expect a publisher to
observe the same injunction. At this period of the
book's production one of their officers assured me
that they knew *Synergetics* was an important book
because they had paid so much for it: top floor and
bottom line were crossing fingers in the increasing
absence of communication with editorial reality.

So it must have been solely on the basis of
Buckminster Fuller's reputation and the knowledge
that this was his most ambitious work that the
corporate genie decreed that in the spring of 1974
Macmillan would herald *Synergetics* as the chief
feature of their fall list, committing the entire front
cover and a full page of copy . . . and this to a book
that virtually no one but Fuller, me, and the copy
editors had ever read! (And how could they? . . . a
book of which most reviewers confess that they
have read only a fraction.)

The price had gone up from a scheduled $12.50
to $15.00. Thirty-two pages of color were promised.
Publication would be in October. The successful
launching of the book seemed assured. Fuller's total
rewriting at white heat had occasioned only a

moderate delay in production, and by April the first bound galleys were finally available.

For some reason the publishing industry finds it cheaper to send the galleys to Barnstable on Cape Cod to be cut and glued into "perfect" bindings than to try to do the same thing at the compositors or in Manhattan. Not until some dozens of these bound galleys had come down from Cape Cod did Macmillan have a realistic opportunity to inspect the merchandise, to see what kind of a product they had on their hands. And the bound galleys were particularly vulnerable as they contained no suggestion of the book's essential artwork—not even the black-and-white line drawings.

As these galleys were routed in-house throughout Macmillan—sales, editorial, legal, promotion—they had to emerge as an accident looking for someplace to happen. Fuller's confrontation with the unbound galleys had provoked a crisis of composition and a crisis of production, but the confrontation of the bound galleys with the real world of management and marketing resulted in catastrophe. A secretary in the publicity department opened the pages to a particularly abstruse passage and complained to her boss that she could understand none of it. The book was instantly judged as totally incomprehensible. Who could possibly understand this kind of talk? *Synergetics* was manifestly a disaster, quite unsuitable for distribution as a trade book in retail stores.

The feeling of shock was not unreasonable, and it took little time for warning lights to flash throughout the house that they had a costly dud on their hands. How could the hard-nosed professionals of Third Avenue have been so taken in by the blandishments of Fuller, strumming his enigmatic refrains as the

wandering minstrel of quantum mechanics? . . . As
our book took on wings, I had seen my role as that
of a stable boy to Pegasus, perhaps, but not as Epios
carving the wooden horse of Troy.

All of this happened at a time when Macmillan
had quite enough other problems on their hands.
They were in the throes of a drastic reorganization
of management and were being harassed with front-
page publicity stemming from deep embroilment in
a labor dispute caused by militant union attempts
to organize the employees. It was a period when no
one's head was on too securely and no one wanted
to lose his in championing a book as crazy as
Synergetics. The officer who had signed our renewed
contract was the first to waver from the cause.
Following the panic in publicity, the people newly
in charge at the trade department disclaimed responsibility and urged dropping the book altogether.

In the summer of 1974, Synergetics was taken
out of production. The book was removed from the
fall list (but it was too late to change the cover). The
paper allotment was canceled. The printing schedule
was scrapped. It was not even tentatively scheduled
for the spring list. The computer read-outs of titles
in production had relegated it to a literary limbo.
And the only way we sensed all this was because no
one would answer the telephone. For Bucky and me
it was like being buried alive.

Our editor, Bruce Carrick, tried vainly—and
charitably—to conceal from us the extent of the
disaster. He later confessed to being guided by his
own dictum that depressed editors are bad enough
but depressed authors are worse. In May, we had
gotten our first warning signal that all was not serene
at the 866 Third Avenue headquarters when Carrick
confided that his superiors were inclined to defer

publication in order to "give the reviewers more time." A month later they said they were considering reprocessing the book as a college textbook. Not until Carrick left Macmillan, in October, did we realize the full extent of the damage. He had kept up a bold front until the last because he felt the arguments against the book had become emotional and that the economics of the situation would argue for eventual publication—if only to cut losses. He was an almost incurable optimist and felt that the company had too much of a financial stake in the project to abandon publication when the chips were down. He had an enthusiastic sense of the market for Fuller's many books, but he was unsuccessful in translating this into in-house support for the project.

During this difficult period of labor disputes and internal administrative turmoil, people were being fired—and rehired—right and left, and it was not hard to get a fairly good picture of the fate of our book from the growing pool of disaffected employees. At times, during that awful summer, it seemed that our only recourse might be litigation over performance on the contract; but none of us wanted to go to court and if we had done that, the book might never have come out.

In retrospect it is the comic, even ludicrous, elements of the picture that survive. A contest of Buckminster Fuller vs. any institution is usually at least an even match; his uncompromising strategy of the individual has a way of winning out and it is usually the institution that ends up in need of consolation and first aid. (Would anyone of sound mind really relish being an editor or literary agent introducing to the real world of New York the kind of uncompromising temperament and original genius

141

that has been compared to Leonardo and Darwin?)
But retrospection was in short supply that summer
and room for introspection was ample. My moustache
went from gray to white and I shaved it. I had to
virtually forego any form of literary collaboration
with Bucky for the better part of a year. We found
we couldn't work together on anything without
getting into the geometry . . . and we couldn't get
into any new geometry without his insistence that it
should be incorporated in the not-yet-printed book.
I knew that whatever prospects the book had would
not be enhanced by our trying to add yet more
material at 59 minutes past the eleventh hour. So
the suspended animation of production entailed a
mindless moratorium on collaboration.

How did the book arouse such violent emotions
before it was even in print? Feelings ran high; it
was a difficult and vulnerable book: 876 pages; 150
illustrations; dozens of tables, charts, and diagrams;
with a deliberately naive and thorny text combining
a homemade psychology and a do-it-yourself
crystallography. One did not have to be a cynic or a
huckster to have reasonable doubts about the
prospects of this kind of work in the marketplace.
I had told Nancy Kahane in Macmillan publicity
that the book was certain to provoke hostile reviews;
I was—perhaps recklessly—trying to suggest that
its controversial character might help it gain
recognition. As it turned out, there were very few
hostile notices, perhaps because reviewers sensibly
preferred not to tangle with a text they could not
master. Most of the first reviewers treated the book
as a gorgeous intellectual toy. It was a book in which
everything—faults and possible virtues—became
larger than life. I speculated that perhaps we had
made a mistake to try to bring it out as a trade

book; maybe we would have been better off with a
technical or specialty publisher—but there the
requisite daring would have been even harder to
find.

Just as Fuller is hard to describe, *Synergetics* is
hard to describe. Just what kind of a book is it?

Synergetics is a book without genre. The Library
of Congress catalogs it succinctly under "1. System
theory. 2. Thought and thinking. 3. Mathematics—
Philosophy." The dilemma of the book is that it
attempts

RBF: It wasn't an attempt. It was spontaneous.

to combine science and poetry and philosophy in a
single work and in the very act of combining three
such elements—normally considered so disparate
in our culture—it is impossible to appeal to any one
of the disciplines without risk of grave offense to
the other two.

I know that the whole structure of Fuller's cosmos
is a poetic one of vast harmony and subtlety. If the
book is nothing else, it is one of the most complex
literary and pattern metaphors of the age. It is a
rare and wonderful vision of a geometry of
conceptuality: how to start from a new place—
independent of Euclid, Descartes, and Leibniz,
independent of size, independent of time. In this
book, Fuller is highly dogmatic, but never mystical;
it is a business of stark homage to the tetrahedron.

*RBF: I am not in homage to anything; certainly not
the tetrahedron as an object, merely as the minimum
structural system in Universe.*

The book expresses a kind of geometry and a
method of epistemology that is probably original in
our civilization; if that is so, its very format means
that academic recognition will be hard to come by.

What I am trying to say is that, while Macmillan did us wrong, they were not without provocation.

Recapitulating our succession of editors over more than a decade at Macmillan is like reciting begats from the Book of Numbers: Peter Ritner, Robert Markel, Samuel Stewart, Ray Roberts, Bruce Carrick, Bernard Hassan, and Michael Denneny. *Synergetics* became an occupational hazard as it gained momentum. Even Ken Shaw, our copy editor, lost his job along with Carrick in the tercimation that wiped out a third of their trade editors. Throughout the fall of 1974—after Carrick was sacked—we had no editor at all. No copy editor either. There was no continuity. Even the secretaries and receptionists were always new faces. (I just thought this was bad for me, not realizing it must have been worse for Macmillan.)

By this time, only by my persistent nosiness, sneaking past the Pinkerton guards in the midst of the strike, could I get some vague notion of what was going on. By poking around at lunchtime one day in late November, I discovered that the *Synergetics* "blues," which had lain idle since April, were gradually being reprocessed. I had no one to communicate with, but some obscure imperative seemed to be at work. Gerry Dickler wrote Macmillan a charming letter inquiring whether the book was to be left "to drift for itself." I finally caught up with the harassed and hard-pressed chief copy editor and chief of production. From them I learned only that all decisions about *Synergetics* were being made on the invisible tenth floor where the top executives were. (Eventually, more than a year later, I was to hear that it was Macmillan's board chairman who broke the impasse on the fifth floor and decreed

that the publication of the book should proceed.
Which, I suppose, just shows that editors don't
have all the fun.)

The book had been designed with 32 pages of
beautiful color drawings—mostly in Fuller's hand—
with intricate color coding for explication of the text.
I learned indirectly that they were considering
running these illustrations in black and white to
reduce costs, although there was no way that the
significance of each illustration could be preserved
without the specific color legend to explain it. The
prospect of dropping the color was a particular
torment for Bucky and prompted his urgent—but in
the end futile—intervention with the top management.
It had been calculated that preserving the color
plates would add about 45 cents in production costs
to each volume. The final decision was to guarantee
the unit profit margin by dropping the color and
raising the retail price to $25.*

I was asked to go through the "blues" and delete
all of the many text references to the color section
and substitute references to what had been lamely
rechristened the "drawings section." The request
struck me as macabre, but there was no one else left
in the company familiar enough with the book to
make the changes. I could not bear to lull the reader
into thinking the pictures were supposed to make
sense in black and white, so I let a few color

*Oddly, this is the only explanation I was given at the time.
Only in the course of writing this book did I find out that
several printers had told the Macmillan production
department that, given the shape the illustrations were in,
it would be impossible to make the color separations by
camera. Macmillan had already spent a fortune on the
preparations of the illustrations; to try to make the color
illustrations work would have meant a staggering addition
to the production costs.

references survive to salve our conscience.

By February the bound books started coming in from Binghamton and Michael Denneny rushed the first copy to Fuller in Philadelphia. It was a Sunday afternoon, and Bucky was in his 31st-floor apartment in I. M. Pei's Society Hill Towers with Isamu Noguchi and Shoji Sadao, when the doorman brought up the special delivery parcel wrapped in brown paper. Bucky had to stop the doorman then and there to tell him about that precious cargo, the book he had been working on all his life. For ten minutes he recapitulated for the astonished doorman how he had first gotten the idea for the book when he was in the Navy, how he had worked with students making the models at so many universities, and the significance of his discoveries for humanity. Then he tipped the doorman ten dollars.

Once Macmillan had remustered its mysterious resolve, they came through handsomely. *Synergetics* was published on 3 April 1975 with a first printing of 15,000 copies. The salesmen had long awaited this title and Denneny had the wit to tell them what the book was about in one word: he said it was a book about the universe.* Macmillan ran generous advertisements in the major city newspapers. (No one having caught a cosmic fish should be expected

---

*Denneny had been anticipated by an unidentified professor at a seminar at Wesleyan University where Fuller was speaking in 1969: he handed Fuller a scribbled note with the following comparison of him and his great-aunt Margaret:

> Fuller is a name
>   For better or for worse,
> Of two who grappled
>   With the universe.
>
> "I'll accept it,"
>   Said the famous spinster.
> "I'll explain it,"
>   Said the bold Buckminster.

to carry it home through an alley.) The president of Macmillan gave a formal press luncheon attended by Raymond Hagel, the Chairman of the Board, and several other corporate officers, for Anne and Bucky and Arthur Loeb and me. Bucky got along famously with their Eastern Division sales manager Frank McCormick, who threw a party for him in Chicago. The *New York Times* and the *Wall Street Journal* featured the book in lengthy reviews with high praise. Macmillan did a great job of getting the book into large and small bookstores all over the country. By fall a second printing of 5,000 copies was ordered, and a retail value of half a million dollars was in print. It took some waiting, but it was quite a package when it finally got out.

The following year on 7 October 1976, Macmillan brought out *And It Came to Pass—But Not to Stay,* a handsome edition of Fuller's accumulated blank verse. As Fuller and I continue to collaborate on the amplification and refinement of his synergetic geometry, we have accumulated over 500 manuscript pages of difficult textual and graphic material comprising *Synergetics 2*—tentatively scheduled for publication as *Synergetics Harvest.* And in the editing of *Cosmic Fishing,* Macmillan has tolerated whatever comments I have wanted to make and granted me generous support and encouragement: their willingness to print my unvarnished remarks accords them the last word which is their due.

11. A Game of Cosmic Solitaire

ONCE *Synergetics* was published, Fuller's feelings
about it were ambivalent. Relief and fulfillment were
diluted by a sense of loss that the bird had flown
the coop. He had completed his ultimate artifact in
the service of humanity, but his temperament was
too restless to admit emeritus satisfactions.

Shortly after he had rewritten the last of the
galleys, he telephoned me from La Jolla, where he
was in a seminar at the Salk Institute, to express an
unfamiliar anxiety. He said that for the first time
he felt disoriented. Consummation was apparently
less congenial than the habitual compulsive race
with destiny—to get it all on paper before he died.
What would happen, now that it was all down on
paper? He was, of course, enormously gratified at
seeing all of his formulas, tables, drawings, and
models in print, but nothing would take the place of
his psychological imperative to recapitulate the
geometric concepts from the beginning, the graphic
process of generating systemic conceptuality out of
stark space-nothingness. Happily, the period when
the first appearance of the book constrained him in
these tendencies was very short-lived.

For me, the chief satisfaction of the book's
publication was at last being able to share the whole
elaborate design with others—though at the price
of having robbed Fuller perhaps of a certain privacy.
His geometry remained difficult, esoteric—even
hermetic—but it was no longer inaccessible, no
longer embedded in untranscribed tapes and scrawls
on the backs of envelopes. I had felt that up until
this time artists and scientists could be excused for

not listening to Fuller; now if they chose to ignore him, it would be on their conscience, not his. Nor mine.

The reaction of many first readers was one of incredulity that Fuller's imagination could function in such total oblivion of the scientific conventions and the cultural traditions of Western civilization. His philosophy was homemade, do-it-yourself, ab initio—as if Pythagoras, Parmenides, Plato, and Aristotle had never existed. It seemed barely possible that Fuller could start from scratch and still tackle the perennial philosophical paradoxes: how to differentiate and relate

—the ideal and the physically realized;

—the container and the contained;

—the one and the many;

—the observer and the observed;

—the human microcosm and the universal macrocosm.

If Fuller ever made use of what the greatest minds had had to say on these subjects, no one ever caught him peeking at their texts. What they had pondered and taught for centuries had become an intrinsic part of our heritage. But not for Fuller; it was not that he was unwilling to bestow them a low bow or even a slight nod, it was just that he felt he had to start from scratch—as he presumed they had done. Had he unlearned what they had taught? Or simply never listened to begin with?

RBF: I don't want any credit for having such wisdom as the ancient philosophers. I was just lucky enough to have been so busy in my youth as an accountant and as a mechanic, never to have learned about them.

When I entered Harvard I had all A's in mathematics and I took some more advanced math, so I was able to catch on to the whole idea of geometric proofs. But

Cosmic Fishing

*I had the good luck to be insulated from all philosophy
and such formal knowledge. They didn't tell you
anything about Plato at Milton Academy. The nearest
we got was the Platonic solids in geometry.*

    *I knew Shakespeare, and the Faerie Queen, and
Thackeray and all about the kings in history: who
beheaded who and who put someone in a tower. I had
the battle stuff okay. . . . But all I knew about the
Greeks was what my mother had taught me, like the
Spartan boy who brought in the fox to eat all his guts.*

How had he come to focus on the same antique
issues if he was so unfamiliar with their writings?
His concerns were simply the coincidental results of
his synergetic strategies, of commencing with wholes
—not parts, not anyone else's building blocks.

    Fuller was not interested in learning about the
great absolutes as derived from what the great minds
before him had to say. He is not anti-cultural; he
just regards being non-cultural as itself the highest
of disciplines.

*RBF: Culture means just getting things stewed up . . .
like growing algae or microbial broth in the laboratory,
keeping the light out.*

    *Culture is purposeless to-ing and fro-ing, back-burner
steering. It is moody drifts, flotsam and jetsam.
Culture is flotsam saying to the jetsam: I think we
ought to have a law against any waves.*

So he had to do it all for himself. He neither affirms
nor rejects Plato's allegory of the Cave; he merely
submits the tetrahedron as the sole and proper
path for its exit—for escape from nonconceptuality
to the first stepping stone of Scenario Universe. And
he observes the Platonic obligation to return to the
darkness of the Cave to describe to the bound
prisoners the dazzling sunlit truth of geometric
reality.

    For me, the book is as much of an enigma as it

ever was. I know that when you create a new
philosophical universe—in your own terms and with
your own rules of play—the resulting edifice may
be logically irrefutable, but not necessarily deserving
of scrutiny. But I have no doubts that Fuller's
scheme merits the most minute examination. (The
substance of the text is his, but certainly it was not
my practice to try to persuade him to popularize or
sugar-coat the language. My rule was to omit no
thorny detail that might conceivably trigger the
imagination of any future student raveling his way
through the labyrinth of mathematical intricacies.)
Only by completely abandoning the static frame of
reference of conventional measurement could Fuller
devise his new starting point for "getting nature in a
corner"—without blackboard, paper, or two lines
crossing. His ambitious goal of integrating all the
disciplines—all of them from sociology and
psychology to electromagnetics and crystallography
—in one grand geometric vision will continue to
outrage, stimulate, and ultimately instruct, many
generations to come.

> RBF: You say it was my ambition to integrate the
> disciplines in one grand system. I did not have any
> ambition about it. I simply assumed that the operation
> of Universe is integrated and you can't understand it
> by isolating any of it or taking it apart.

With its finite discontinuities, its three-way great
circle grid, its 60-degree coordination, and its
tetrahedral matrix, synergetics mathematics suggests
many analogies and correspondences to recent
observations in the physical sciences. Viral
structures reveal themselves as geodesic. Geological
studies in plate tectonics invite polyhedral
analysis. Ancient monuments enshrine ratios from
closest-sphere-packing hierarchies. Such vistas are

151

always tantalizing, and sometimes exhilarating. Even in the absence of such lofty confirmation, Fuller's geometry remains a towering intellectual achievement.

As any devotee of Fuller's lectures knows, the great spiritual climacteric of his life was his crisis of 1927 when he was living in Chicago at a level of bare subsistence, when he resolved not to utter a single word until he could define it to his complete satisfaction. The result was many months when he did not speak at all. His speech today is now purged of many words from his earlier vocabulary—our common social idiom—which did not survive the testing of that period. This rendezvous with first principles and retreat into silence was an episode of intense trial, withdrawal, and torment. His listeners are vicariously compelled to share the lonely experience of his self-examination, a metaphor for the source of his commitment to humanity. Somewhere in the course of this transcendent spiritual crisis it is certain that Fuller visited a past or future landscape of astronomically remote philosophic distance.

It has always been my suspicion that some part of him remains in that alien country of his self-discovery, that he has never fully returned. During the long months of silence, his continuity of mundane physical existence was so tenuous that he seems barely to have survived the re-entry to fulfill his allotted life span on earth. This alien passage is the source of his loneliness and of his fractured sense of identity. (In addition to his office and home telephone numbers Fuller has always kept—in Carbondale or in Philadelphia—an unlisted telephone number. Conversely, over these same decades he

has maintained an overt listing in New York's
Manhattan telephone directory linked to an
answering service in Flushing. He is obsessed with
both self-isolation and the need to leave out the
latchstring for some wayward stranger.) Here is a
man who invariably shares with his lecture audiences
the disturbing thought that no man can prove upon
awakening that he is the man who he thinks went
earlier to sleep. He still yearns to reduce the sense
of loss and isolation. This is why he responds so to
the innocence of children. This is why he feels so
at home in metaphysical companionship with the
yogi.* Being alone was the original price; loved, he
says, but alone.

Fuller has to keep constantly reminding himself
that he has completed the journey back to earth
from the intellectual orbit; unlike other mortals who
have ventured less far, he has to keep pinching
himself for reassurance that he has returned to, and
still inhabits, the familiar physical persona. (When
I asked Bucky what it was like to come back to
earth, he confirmed my description and confided to
me that all the good part of him—"all that my
mother would like"—came back from orbit and only
the bad part was left behind.)

Here was a bourn from which not only had
traveler returned, but returned with a kind of cosmic
zip code by which he could continue to receive
messages. Here, at least, was a way of accounting
for a recurring phenomenon in the course of writing
Synergetics, when Bucky would say that he felt as
if he were an *agent* for some transcendent or
supernatural source of inspiration, as if he were

*Fuller is entranced by yogis, but censures them for using
their powers for self-development rather than for the service
of others.

merely an interceptor or transceiver of messages originating elsewhere. It was not like a trance or automatic writing, but there were many occasions when he could not provide a rational accounting for what he had just written or said—nor did he even pretend to recognize its full significance. The *thought* had its own integrity independent of the thinker, and we respected it accordingly. "The thought," he says, "does not belong to you."

Of course, Fuller has always maintained that he does not invent his thoughts, that he merely separates out some local patterns from a confusing whole—confusing just meaning "untuned" or unfused. In his world, thoughts are a priori, having a reality independent of, and antecedent to, the thinker; the individual thinker becomes merely their vessel. He has always suggested that our intuitive thoughts may be simply remote cosmic transmissions. This is how he can play his game of solitaire without the hint of arrogance. Solitaire is a game that anyone can play; that no one else appears to have played this particular kind of game is regrettable, but he is sure others will learn.

Twenty-three years after the Chicago crisis of 1927, Fuller was addressing a group of design students in Raleigh, North Carolina, and told them

Energetic geometry is a game of solitaire I started playing in 1917. . . . I assume that what I began to discover in the game was territory that had been well covered by students long ago. In some instances that would seem to be the case. In other instances it would seem that we have rediscovered arrangements of phenomena that, if they have been known to man, have long since been forgotten.

I must say that when you play a game like this you get a strange feeling when you come into view of arrangements of components of your energetic Universe with which you are not at all familiar and which you are quite sure men have not seen recently. Yet you have the astounding feeling that someone was here only seven thousand years ago, or something like that. You get the feeling of a close kinship to the intellectual speculation of all time. You sometimes feel that this time you can make it stick.

When Fuller first saw the great hexagonal court in the ancient ruins of Baalbek, he said, "The Phoenicians knew my principles." Broadening the theme he wrote in *Intuition* (1970)

Certain it is on my own part
That I have made several mathematical discoveries
Of a fundamental unexpected and unpublished nature.
As I realized my discovery
I always have had
The same strange sensation
That this newly realized conception
Previously unknown to terrestrial humans,
Had been known
To the human mind
Sometime vastly long ago.

Thus his mathematical discoveries may have been known since the distant past, but not necessarily on earth. He dismisses the conventional accounts of the history of scientific discoveries on the grounds that they are too limited to the narrow and peculiar circumstances of Western civilization. All of the knowledge in the universe may have been known to various people at various other times. "However distant or remote any information signal is, it has to just go on forever unless it is intercepted. I look upon myself as an agent," but, he adds significantly, "all of us are."

RBF: *The information in Universe has always been there. The total information is always there, but it has been deployed into generalized-principle increments of cosmic tunability.*

Cosmic Fishing

The information signals are forever bouncing electromagnetically about the Universe, every so often impinging on celestial entities and being either tunably received or bounced off to travel elsewhere.

If we fail to catch a cosmic fish it may be a trillion years before the opportunity comes again. It will come . . . but it may not be in this Galaxy. Sumtotally, all the fish will always eventually be caught and rebroadcast, but not all at the same rebroadcasting stations.

Acknowledgments

There are two ways to write a book: show it to nobody or show it to everybody. This book has greatly benefited from the latter course and I am indebted in varying degrees to June, Jarratt, Ashton, Maria, and Anthony Applewhite; Lawrence M. Connell; Isobel and John Fistere; Anne Hewlett Fuller; Susan Griggs; Mary Anne Kenner; Arthur L. Loeb; Eugene Milligan; Frances and John G. Shaffer; Ken Shaw; William M. Wainwright; Nancy Z. Walworth; and Louise Applewhite Watson.

At Macmillan, I am in particular debt to Michael Denneny, an editor's editor; his assistant, Beth Rashbaum; and to Maurice Schneps.

3 ♂